Textile Engineering: Materials, Design and Technology

Textile Engineering: Materials, Design and Technology

Editor: Dominic Tate

NY RESEARCH
P R E S S

New York

Published by NY Research Press
118-35 Queens Blvd., Suite 400,
Forest Hills, NY 11375, USA
www.nyresearchpress.com

Textile Engineering: Materials, Design and Technology
Edited by Dominic Tate

International Standard Book Number: 978-1-63238-597-0 (Hardback)

This book contains information obtained from authentic and highly regarded sources. All chapters are published with permission under the Creative Commons Attribution Share Alike License or equivalent. A wide variety of references are listed. Permissions and sources are indicated; for detailed attributions, please refer to the permissions page. Reasonable efforts have been made to publish reliable data and information, but the authors, editors and publisher cannot assume any responsibility for the vailidity of all materials or the consequences of their use.

Trademark Notice: Registered trademark of products or corporate names are used only for explanation and identification without intent to infringe.

Cataloging-in-Publication Data

Textile engineering : materials, design and technology / edited by Dominic Tate.
     p. cm.
Includes bibliographical references and index.
ISBN 978-1-63238-597-0
1. Textile fabrics--Technological innovations. 2. Textile industry--Technological innovations.
3. Textile design. 4. Textile machinery. I. Tate, Dominic.
TS1765 .T49 2018
677--dc23

Contents

Permissions

Index

Preface

Textile manufacturing refers to the process of converting cotton, silk, jute, etc. into fabric to be used for clothes. The process primarily starts with knitting and weaving. Some other forms are braiding, plaiting, and bonding of fibers. The most common types of textiles manufactured around the world are linen, wool, cashmere, velvet, rayon, denim, etc. This book is a compilation of chapters that discuss the most vital concepts in the field of textile manufacturing. It explores all the important aspects of the field in the present day scenario. This textbook attempts to assist those with a goal of delving into the field of textile manufacturing.

To facilitate a deeper understanding of the contents of this book a short introduction of every chapter is written below:

Chapter 1- Textile manufacturing is an industry related to the creation of fabric. These fabrics are printed or dyed and then converted into clothes. Handmade techniques are still very popular in this industry. The chapter on fabric manufacturing offers an insightful focus, keeping in mind the complex subject matter.

Chapter 2- Weaving and knitting are techniques used in textile production. Weaving is a repetition of mainly three actions, shedding, picking and battening. The common types of weaves are satin weave, plain weave and twill. This chapter discusses the processes and technology related to weaving and knitting in a critical manner providing key analysis to the subject matter.

Chapter 3- Design is very important for the aesthetic value of the cloth. Fabric weave design means the pattern that exists between the warp and weft yarns. Bringing changes in the drafting and lifting plan helps in creating a design. Textile weaving design is best understood in confluence with the major topics listed in the following chapter.

Chapter 4- Nonwovens are fabrics that are made from long and staple fibers. The technology used in these fabrics is different from weaving and knitting. Nonwoven fabrics are bonded together by heat, chemical or with solvent treatment. The topics discussed in the chapter are of great importance to broaden the existing knowledge on nonwoven fabrics.

Chapter 5- Winding as a process can be explained as the transfer of spinning yarn from a package to another. There are different types of winding principles such as spindle-driven winders, pirn winders, yarn tensioning and splicing. The major categories of winding are dealt with great details in the chapter.

Chapter 6- The warping process converts the yarn into a usable product which has the requisite number of ends, width and length. Warping is usually done on a frame or loom.

Size coating of yarn prevents wear and tear by applying a uniform amount of yarn surface on a loom or beam. The topics discussed in the chapter are of great importance to broaden the existing knowledge on fabric manufacture.

I owe the completion of this book to the never-ending support of my family, who supported me throughout the project.

Editor

An Overview of Textile Manufacturing

Textile manufacturing is an industry related to the creation of fabric. These fabrics are printed or dyed and then converted into clothes. Handmade techniques are still very popular in this industry. The chapter on fabric manufacturing offers an insightful focus, keeping in mind the complex subject matter.

Textile Manufacturing

Textile manufacturing is a major industry. It is based on the conversion of fibre into yarn, yarn into fabric. These are then dyed or printed, fabricated into clothes. Different types of fibre are used to produce yarn. Cotton remains the most important natural fibre, so is treated in depth. There are many variable processes available at the spinning and fabric-forming stages coupled with the complexities of the finishing and colouration processes to the production of a wide ranges of products. There remains a large industry that uses hand techniques to achieve the same results.

Processing of Cotton

Cotton is the world's most important natural fibre. In the year 2007, the global yield was 25 million tons from 35 million hectares cultivated in more than 50 countries.

There are six stages

- Cultivating and Harvesting
- Preparatory Processes
- Spinning
- Weaving or Knitting
- Finishing
- Marketing

Cultivating and Harvesting

Cotton is grown anywhere with long, hot dry summers with plenty of sunshine and low humidity. Indian cotton, gossypium arboreum, is finer but the staple is only suitable for hand processing. American cotton, gossypium hirsutum, produces the longer staple needed for machine production. Planting is from September to mid November and the crop is harvested between March and June. The cotton bolls are harvested by stripper harvesters and spindle pickers, that remove the entire boll from the plant. The cotton boll is the seed pod of the cotton plant, attached to each of the thousands of seeds are fibres about 2.5 cm long.

- Ginning

    The seed cotton goes into a Cotton gin. The cotton gin separates seeds and removes the "trash" (dirt, stems and leaves) from the fibre. In a saw gin, circular saws grab the fibre and pull it through a grating that is too narrow for the seeds to pass. A roller gin is used with longer staple cotton. Here a leather roller captures the cotton. A knife blade, set close to the roller, detaches the seeds by drawing them through teeth in circular saws and revolving brushes which clean them away.

    The ginned cotton fibre, known as lint, is then compressed into bales which are about 1.5 m tall and weigh almost 220 kg. Only 33% of the crop is usable lint. Commercial cotton is priced by quality, and that broadly relates to the average length of the staple, and the variety of the plant. Longer staple cotton (2½ in to 1¼ in) is called Egyptian, medium staple (1¼ in to ¾ in) is called American upland and short staple (less than ¾ in) is called Indian.

    The cotton seed is pressed into a cooking oil. The husks and meal are processed into animal feed, and the stems into paper.

Preparatory Processes - Preparation of Yarn

- Ginning, bale-making and transportation is done in the country of origin.

- Opening and cleaning

Platt Bros. Picker

Cotton mills get the cotton shipped to them in large, 500 pound bales. When the cotton comes out of a bale, it is all packed together and still contains vegetable matter. The bale is broken open using a machine with large spikes. It is called an Opener. In order to fluff up the cotton and remove the vegetable matter, the cotton is sent through a picker, or similar machines. The cotton is fed into a machine known as a picker, and gets beaten with a beater bar in order to loosen it up. It is fed through various rollers, which serve to remove the vegetable matter. The cotton, aided by fans, then collects on a screen and gets fed through more rollers till it emerges as a continuous soft fleecy sheet, known as a lap.

- Blending,

  Mixing and Scutching

Scutching refers to the process of cleaning cotton of its seeds and other impurities. The first scutching machine was invented in 1797, but did not come into further mainstream use until after 1808 or 1809, when it was introduced and used in Manchester, England. By 1816, it had become generally adopted. The scutching machine worked by passing the cotton through a pair of rollers, and then striking it with iron or steel bars called beater bars or beaters. The beaters, which turn very quickly, strike the cotton hard and knock the seeds out. This process is done over a series of parallel bars so as to allow the seeds to fall through. At the same time, air is blown across the bars, which carries the cotton into a cotton chamber.

- Carding

Carding machine   A Combing machine

Carding: the fibres are separated and then assembled into a loose strand (sliver or tow) at the conclusion of this stage.

The cotton comes off of the picking machine in laps, and is then taken to carding machines. The carders line up the fibres nicely to make them easier to spin. The carding machine consists mainly of one big roller with smaller ones surrounding it. All of the rollers are covered in small teeth, and as the cotton progresses further on the teeth get finer (i.e. closer together). The cotton leaves the carding machine in the form of a sliver; a large rope of fibres.

Note: In a wider sense Carding can refer to these four processes: Willowing-loosening the fibres; Lapping- removing the dust to create a flat sheet or lap of cotton; Carding- combing the tangled lap into a thick rope of 1/2 in in diameter, a sliver; and Drawing- where a drawing frame combines 4 slivers into one- repeated for increased quality.

- Combing is optional, but is used to remove the shorter fibres, creating a stronger yarn.

- Drawing the fibres are straightened

Several slivers are combined. Each sliver will have thin and thick spots, and by combining several slivers together a more consistent size can be reached. Since combining several slivers produces a very thick rope of cotton fibres, directly after being combined the slivers are separated into rovings. These rovings (or slubbings) are then what are used in the spinning process.

Generally speaking, for machine processing, a roving is about the width of a pencil.

- Drawing frame: Draws the strand out

- Slubbing Frame: adds twist, and winds onto bobbins

- Intermediate Frames: are used to repeat the slubbing process to produce a finer yarn.

- Roving frames: reduces to a finer thread, gives more twist, makes more regular and even in thickness, and winds onto a smaller tube.

Spinning - Yarn Manufacture

- Spinning

Most spinning today is done using Break or Open-end spinning, this is a technique where the staples are blown by air into a rotating drum, where they attach themselves to the tail of formed yarn that is continually being drawn out of the chamber. Other methods of break spinning use needles and electrostatic forces. This method has replaced the older methods of ring and mule spinning. It also is easily adapted for artificial fibres.

The spinning machines takes the roving, thins it and twists it, creating yarn which it winds onto a bobbin.

In mule spinning the roving is pulled off a bobbin and fed through some rollers, which are feeding at several different speeds. This thins the roving at a consistent rate. If the roving was not a consistent size, then this step could cause a break in the yarn, or could jam the machine. The yarn is twisted through the spinning of the bobbin as the carriage moves out, and is rolled onto a cylinder called a spindle, which then produces a cone-shaped bundle of fibres known as a "cop", as the carriage returns. Mule spinning produces a finer thread than the less skilled ring spinning.

- The mule was an intermittent process, as the frame advanced and returned a distance of 5ft.It was the descendant of 1779 Crompton device. It produces a softer less twisted thread that was favoured for fines and for weft.

- The ring was a descendant of the Arkwright water Frame 1769. It was a continuous process, the yarn was coarser, had a greater twist and was stronger so was suited to be warp. Ring spinning is slow due to the distance the thread must pass around the ring, other methods have been introduced.

Sewing thread, was made of several threads twisted together, or doubled.

- Checking.

This is the process where each of the bobbins is rewound to give a tighter bobbin.

- Folding and twisting

Plying is done by pulling yarn from two or more bobbins and twisting it together, in the opposite direction that in which it was spun. Depending on the weight desired, the cotton may or may not be plied, and the number of strands twisted together varies.

- Gassing

Gassing is the process of passing yarn, as distinct from fabric very rapidly through a series of Bunsen gas flames in a gassing frame, in order to burn off the projecting fibres and make the thread round and smooth and also brighter. Only the better qualities of yarn are gassed, such as that used for voiles, poplins, venetians, gabardines, many Egyptian qualities, etc. There is a loss of weight in gassing, which varies' about 5 to 8 per cent., so that if a 2/60's yarn is required 2/56's would be used. The gassed yarn is darker in shade afterwards, but should not be scorched.

Mule spinning

Mule spinning

Ring spinning

Ring spinning

Measurements

- Cotton Counts: Refers to the thickness of the cotton yarn where 840 yards of yarns weighs 1 pound (0.45 kg). 10 count cotton means that 8,400 yards (7,700 m) of yarn weighs 1 pound (0.45 kg). This is coarser than 40 count cotton where 40x840 yards are needed. In the United Kingdom, Counts to 40s are coarse (Oldham Counts), 40 to 80s are medium counts and above 80 is a fine count. In the United States ones to 20s are coarse counts.

- Hank: A length of 7 leas or 840 yards.

- Thread: A length of 54 in (the circumference of a warp beam).

- Bundle: Usually 10 lb.

- Lea: A length of 80 threads or 120 yards.

- Denier: this is an alternative method. It is defined as a number that is equivalent to the weight in grams of 9000m of a single yarn. 15 denier is finer than 30 denier.

- Tex: is the weight in grams of 1 km of yarn.

The worsted hank is only 560 yd.

Weaving-fabric Manufacture

The weaving process uses a loom. The lengthway threads are known as the warp, and the cross way threads are known as the weft. The warp which must be strong needs to be presented to loom on a warp beam. The weft passes across the loom in a shuttle, that carries the yarn on a pirn. These pirns are automatically changed by the loom. Thus, the yarn needs to be wrapped onto a beam, and onto pirns before weaving can commence.

- Winding

  After being spun and plied, the cotton thread is taken to a warping room where the winding machine takes the required length of yarn and winds it onto warpers bobbins.

- Warping or beaming

A Warper

  Racks of bobbins are set up to hold the thread while it is rolled onto the warp bar of a loom. Because the thread is fine, often three of these would be combined to get the desired thread count.

- Sizing

  Slasher sizing machine needed for strengthening the warp by adding starch to reduce breakage of the yarns.

- Drawing in, Looming

  The process of drawing each end of the warp separately through the dents of the reed and the eyes of the healds, in the order indicated by the draft.

- Pirning (Processing the weft)

  Pirn winding frame was used to transfer the weft from cheeses of yarn onto the pirns that would fit into the shuttle.

- Weaving

At this point, the thread is woven. Depending on the era, one person could manage anywhere from 3 to 100 machines. In the mid nineteenth century, four was the standard number. A skilled weaver in 1925 would run 6 Lancashire Looms. As time progressed new mechanisms were added that stopped the loom any time something went wrong. The mechanisms checked for such things as a broken warp thread, broken weft thread, the shuttle going straight across, and if the shuttle was empty. Forty of these Northrop Looms or automatic looms could be operated by one skilled worker.

A Draper loom in textile museum, Lowell, Massachusetts

The three primary movements of a loom are shedding, picking, and beating-up.

- *Shedding*: The operation of dividing the warp into two lines, so that the shuttle can pass between these lines. There are two general kinds of sheds-"open" and "closed." Open Shed-The warp threads are moved when the pattern requires it-from one line to the other. Closed Shed-The warp threads are all placed level in one line after each pick.

- *Picking*:The operation of projecting the shuttle from side to side of the loom through the division in the warp threads. This is done by the over-pick or underpick motions. The overpick is suitable for quick-running looms, whereas the underpick is best for heavy or slow looms.

- *Beating-up*: The third primary movement of the loom when making cloth, and is the action of the reed as it drives each pick of weft to the fell of the cloth.

The Lancashire Loom was the first semi-automatic loom. Jacquard looms and Dobby looms are looms that have sophisticated methods of shedding. They may be separate looms, or mechanisms added to a plain loom. A Northrop Loom was fully automatic and was mass produced between 1909 and the mid-1960s. Modern looms run faster and do not use a shuttle: there are air jet looms, water jet looms and rapier looms.

Measurements

- Ends and Picks: Picks refer to the weft, ends refer to the warp. The coarseness of the cloth can be expressed as the number of picks and ends per quarter inch square, or per inch square. Ends is always written first. For example: *Heavy domestics are made from coarse yarns, such as 10's to 14's warp and weft, and about 48 ends and 52 picks.*

Associated Job Titles

- Piecer
- Scavenger
- Weaver
- Tackler
- Draw boy
- Pirner

Issues

When a hand loom was located in the home, children helped with the weaving process from an early age. Piecing needs dexterity, and a child can be as productive as an adult. When weaving moves from the home to the mill, children are often allowed to *help* their older sisters, and laws have to be made to prevent child labour becoming established.

Knitting- Fabric Manufacture

Knitting by machine is done in two different ways; warp and weft. Weft knitting is similar in method to hand knitting with stitches all connected to each other horizontally. Various weft machines can be configured to produce textiles from a single spool of yarn or multiple spools depending on the size of the machine cylinder (where the needles are bedded). In a warp knit there are many pieces of yarn and there are vertical chains, zigzagged together by crossing the Cotton yarn.

A circular knitting machine.

Close-up on the needles.

Warp knits do not stretch as much as a weft knit, and it is run-resistant. A weft knit is not run-resistant, but stretches more. This is especially true if spools of spandex are processed from separate spool containers and interwoven through the cylinder with cotton yarn, giving the finished product more flexibility and making it less prone to having a 'baggy' appearance. The average t-shirt is a weft knit.

Finishing- Processing of Textiles

The woven cotton fabric in its loom-state not only contains impurities, including warp size, but requires further treatment in order to develop its full textile potential. Furthermore, it may receive considerable added value by applying one or more finishing processes.

- Desizing

   Depending on the size that has been used, the cloth may be steeped in a dilute acid and then rinsed, or enzymes may be used to break down the size.

- Scouring

   Scouring, is a chemical washing process carried out on cotton fabric to remove natural wax and non-fibrous impurities (e.g. the remains of seed fragments) from the fibres and any added soiling or dirt. Scouring is usually carried in iron vessels called kiers. The fabric is boiled in an alkali, which forms a soap with free fatty acids (saponification). A kier is usually enclosed, so the solution of sodium hydroxide can be boiled under pressure, excluding oxygen which would degrade the cellulose in the fibre. If the appropriate reagents are used, scouring will also remove size from the fabric although desizing often precedes scouring and is considered to be a separate process known as fabric preparation. Preparation and scouring are prerequisites to most of the other finishing processes.

At this stage even the most naturally white cotton fibres are yellowish, and bleaching, the next process, is required.

- Bleaching

Bleaching improves whiteness by removing natural coloration and remaining trace impurities from the cotton; the degree of bleaching necessary is determined by the required whiteness and absorbency. Cotton being a vegetable fibre will be bleached using an oxidizing agent, such as dilute sodium hypochlorite or dilute hydrogen peroxide. If the fabric is to be dyed a deep shade, then lower levels of bleaching are acceptable, for example. However, for white bed sheetings and medical applications, the highest levels of whiteness and absorbency are essential.

- Mercerising

A further possibility is mercerizing during which the fabric is treated with caustic soda solution to cause swelling of the fibres. This results in improved lustre, strength and dye affinity. Cotton is mercerized under tension, and all alkali must be washed out before the tension is released or shrinkage will take place. Mercerizing can take place directly on grey cloth, or after bleaching.

Many other chemical treatments may be applied to cotton fabrics to produce low flammability, crease resist and other special effects but four important non-chemical finishing treatments are:

- Singeing

Singeing is designed to burn off the surface fibres from the fabric to produce smoothness. The fabric passes over brushes to raise the fibres, then passes over a plate heated by gas flames.

- Raising

Another finishing process is raising. During raising, the fabric surface is treated with sharp teeth to lift the surface fibres, thereby imparting hairiness, softness and warmth, as in flannelette.

- Calendering

Calendering is the third important mechanical process, in which the fabric is passed between heated rollers to generate smooth, polished or embossed effects depending on roller surface properties and relative speeds.

- Shrinking (Sanforizing)

Finally, mechanical shrinking (sometimes referred to as sanforizing), whereby the fabric is forced to shrink width and/or lengthwise, creates a fabric in which any residual tendency to shrink after subsequent laundering is minimal.

- Dyeing

  Finally, cotton is an absorbent fibre which responds readily to colouration processes. Dyeing, for instance, is commonly carried out with an anionic direct dye by completely immersing the fabric (or yarn) in an aqueous dyebath according to a prescribed procedure. For improved fastness to washing, rubbing and light, other dyes such as vats and reactives are commonly used. These require more complex chemistry during processing and are thus more expensive to apply.

- Printing

  Printing, on the other hand, is the application of colour in the form of a paste or ink to the surface of a fabric, in a predetermined pattern. It may be considered as localised dyeing. Printing designs onto already dyed fabric is also possible.

Economic, Environmental and Political Consequences of Cotton Manufacture

The growth of cotton is divided into two segments i.e. organic and genetically modified. Cotton crop provides livelihood to millions of people but its production is becoming expensive because of high water consumption, use of expensive pesticides, insecticides and fertiliser. Genetically Modified products aim to increase disease resistance and reduce the water required. The organic sector was worth $583 million. Genetically Modified cotton, in 2007, occupied 43% of cotton growing areas.

Cotton is farmed intensively and uses large amounts of fertilizer and 25% of the world's insecticides. Native Indian varieties of cotton were rainwater fed, but modern hybrids used for the mills need irrigation, which spreads pests. The 5% of cotton-bearing land in India uses 55% of all pesticides used in India. In United Kingdom some companies design cloths for manufacturers such as Sewport, and Bridge & Stitch.

The consumption of energy in form of water and electricity is relatively high, especially in processes like washing, de-sizing, bleaching, rinsing, dyeing, printing, coating and finishing. Processing is time consuming. The major portion of water in textile industry is used for wet processing of textile (70 per cent). Approximately 25 per cent of energy in the total textile production like fibre production, spinning, twisting, weaving, knitting, clothing manufacturing etc. is used in dyeing. About 34 per cent of energy is consumed in spinning, 23 per cent in weaving, 38 per cent in chemical wet processing and five per cent in miscellaneous processes. Power dominates consumption pattern in spinning and weaving, while thermal energy is the major factor for chemical wet processing.

Before mechanisation, cotton was harvested manually by farmers in India and by African slaves in America. In 2012 Uzbekistan was a major exporter of cotton and uses manual labour during the harvest. Human rights groups claim that health care professionals and children are forced to pick cotton.

Processing of other Vegetable Fibres

Flax

Flax is a bast fibre, which means it comes in bundles under the bark of the Linum usitatissimum plant. The plant flowers and is harvested.

- Retting
- Breaking
- Scutching
- Hackling or combing

It is now treated like cotton.

Jute

Jute is a bast fibre, which comes from the inner bark of the plants of the Corchorus genus. It is retted like flax, sundried and baled. When spinning a small amount of oil must be added to the fibre. It can be bleached and dyed. It was used for sacks and bags but is now used for the backing for carpets. Jute can be blended with other fibres to make composite fabrics and work continues in Bangladesh to refine the processes and extend the range of usage possible. In the 1970s, jute-cotton composite fabrics were known as *jutton* fabrics.

Hemp

Hemp is a bast fibre from the inner bark of Cannabis sativa. It is difficult to bleach, it is used for making cord and rope.

- Retting
- Separating
- Pounding

Other Bast Fibres

These bast fibres can also be used: kenaf, urena, ramie, nettle.

Other Leaf Fibres

Sisal is the main leaf fibre used; others are: abacá and henequen.

Processing of Protein Fibres

Wool

Wool comes from domesticated sheep. It forms two products, woolens and worsteds. The sheep has two sorts of wool and it is the inner coat that is used. This can be mixed with wool that has been recovered from rags. Shoddy is the term for recovered wool that is not matted, while mungo comes from felted wool. Extract is recovered chemically from mixed cotton/wool fabrics.

The fleece is cut in one piece from the sheep.This is then skirted to remove the soiled wool, and baled. It is graded into long wool where the fibres can be up to 15 in, but anything over 2.5 inches is suitable for combing into worsteds. Fibres less than that form short wool and are described as clothing or carding wool.

At the mill the wool is scoured in a detergent to remove grease (the yolk) and impurities. This is done mechanically in the opening machine. Vegetable matter can be removed chemically using sulphuric acid (carbonising). Washing uses a solution of soap and sodium carbonate. The wool is oiled before carding or combing.

- Woollens: Use noils from the worsted combs, mungo and shoddy and new short wool

- Worsteds

    Combing: Oiled slivers are wound into laps, and placed in the circular comber. The worsted yarn gathers together to form a top. The shorter fibres or noils remain behind and are removed with a knife.

- Angora

Silk

The processes in silk production are similar to those of cotton but take account that reeled silk is a continuous fibre. The terms used are different.

- Opening bales. Assorting skeins: where silk is sorted by colour, size and quality, scouring: where the silk is washed in water of 40 degrees for 12 hours to remove the natural gum, drying: either by steam heating or centrifuge, softening: by rubbing to remove any remaining hard spots.

- Silk throwing (winding). The skeins are placed on a reel in a frame with many others. The silk is wound onto spools or bobbins.

    - Doubling and twisting. The silk is far too fine to be woven, so now it is doubled and twisted to make the warp, known as organzine and the weft, known as tram. In organzine each single is given a few twists per inch (tpi), and combine with several other singles counter twisted hard at 10 to 14 tpi. In tram the two singles are doubled with each other with a light twist, 3 to 6 tpi. Sewing thread is two tram threads, hard twisted, and machine-twist is made of three hard-twisted tram threads. Tram for the crepe process is twisted at up to 80 tpi to make it 'kick up'.

- Stretching. The thread is tested for consistent size. Any uneven thickness is stretched out. The resulting thread is reeled into containing 500 yd to 2500 yd. The skeins are about 50 in in loop length.

- Dyeing: the skeins are scoured again, and discoloration removed with a sulphur process. This weakens the silk. The skeins are now tinted or dyed. They are dried and rewound onto bobbins, spools and skeins. Looming, and the weaving process on power looms is the same as with cotton.

- Weaving. The organzine is now warped. This is a similar process to in cotton. Firstly, thirty threads or so are wound onto a warping reel, and then using the warping reels, the threads are beamed. A thick layer of paper is laid between each layer on the beam to stop entangling.

Processing of Synthetic Fibres

Discussion of Types of Synthetic Fibers

Synthetic fibers are the result of extensive development by scientists to improve upon the naturally occurring animal and plant fibers. In general, synthetic fibers are created by forcing, or extruding, fiber forming materials through holes (called spinnerets) into the air, thus forming a thread. Before synthetic fibers were developed, cellulose fibers were made from natural cellulose, which comes from plants.

The first artificial fiber, known as art silk from 1799 onwards, became known as viscose around 1894, and finally rayon in 1924. A similar product known as cellulose acetate was discovered in 1865. Rayon and acetate are both artificial fibers, but not truly synthetic, being made from wood. Although these artificial fibers were discovered in the mid-nineteenth century, successful modern manufacture began much later in the 1930s. Nylon, the first synthetic fiber, made its debut in the United States as a replacement for silk, and was used for parachutes and other military uses.

The techniques used to process these fibers in yarn are essentially the same as with natural fibers, modifications have to be made as these fibers are of great length, and have no texture such as the scales in cotton and wool that aid meshing.

Fabric Manufacturing Technologies

Textile fabrics are generally two dimensional flexible materials made by interlacing of yarns or inter-meshing of loops with the exception of nonwovens and braids. Fabric manufacturing is one of the four major stages (fibre production, yarn manufacturing, fabric manufacturing, and textile chemical processing) of textile value chain. Most of the apparel fabrics are manufactured by weaving technology though knitting is catching up fast specially in the sportswear segment. Natural fibres in general and cotton fibre in particular are the most popular raw material for woven fabrics intended for apparel use. Staple fibres are converted into spun yarns by the use of a series of machines in the yarn manufacturing section. Continuous filament yarns are texturised to impart spun yarn like bulk and appearance to them.

Textile fabrics are special materials as they are generally light-weight, flexible (easy to bend, shear and twist), moldable, permeable and strong. There are four major technologies of fabric manufacturing as listed below.

- Weaving

- Knitting

- Non-woven

- Braiding

   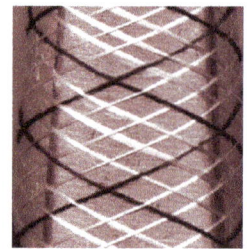

Woven   Knitted   Nonwoven   Braided
Fabrics produced by different technologies

Fabric manufacturing may be preceded either by fibre production (in case of nonwoven) or by yarn manufacturing (in case of weaving, knitting and braiding). Fabrics intended for apparel use must fulfill multidimensional quality requirements in terms of drape, handle, crease recovery, tear strength, air permeability, thermal resistance, moisture vapour permeability. However, looking at the unique properties and versatility of textile fabrics, they are now being used in various technical applications where the requirements are altogether different. Some examples are given in table.

Table: Properties of some technical fabrics

Fabric type	Important properties/ parameters
Filter fabric	Pore size, pore size distribution
Body armour fabrics	Impact resistance, areal density, bending resistance
Fabrics as performs for composite	Tensile strength and tensile modulus
Knitted compression bandages	Stretchability, tensile modulus, creep

Weaving and Knitting: Textile Production Methods

Weaving and knitting are techniques used in textile production. Weaving is a repetition of mainly three actions, shedding, picking and battening. The common types of weaves are satin weave, plain weave and twill. This chapter discusses the processes and technology related to weaving and knitting in a critical manner providing key analysis to the subject matter.

Weaving

Weaving is a method of textile production in which two distinct sets of yarns or threads are interlaced at right angles to form a fabric or cloth. Other methods are knitting, felting, and braiding or plaiting. The longitudinal threads are called the warp and the lateral threads are the weft or filling. (*Weft* or *woof* is an old English word meaning "that which is woven".) The method in which these threads are inter woven affects the characteristics of the cloth. Cloth is usually woven on a loom, a device that holds the warp threads in place while filling threads are woven through them. A fabric band which meets this definition of cloth (warp threads with a weft thread winding between) can also be made using other methods, including tablet weaving, back-strap, or other techniques without looms.

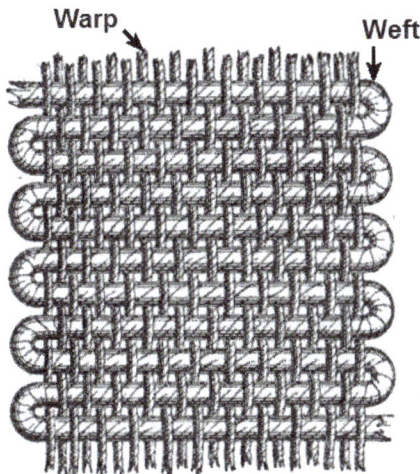

Warp and weft in plain weaving

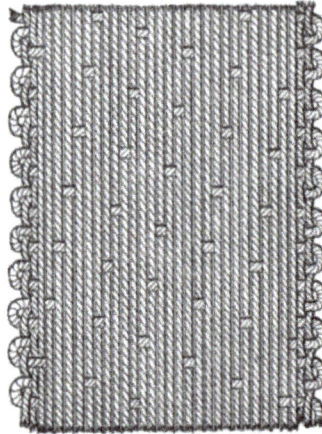

A satin weave, common for silk, each warp thread floats over 16 weft threads.

The way the warp and filling threads interlace with each other is called the weave. The majority of woven products are created with one of three basic weaves: plain weave, satin weave, or twill. Woven cloth can be plain (in one colour or a simple pattern), or can be woven in decorative or artistic design.

Process and Terminology

In general, weaving involves using a loom to interlace two sets of threads at right angles to each other: the warp which runs longitudinally and the weft (older *woof*) that crosses it. One warp thread is called an end and one weft thread is called a pick. The warp threads are held taut and in parallel to each other, typically in a loom. There are many types of looms.

Weaving can be summarized as a repetition of these three actions, also called the primary motion of the loom.

- Shedding: where the ends are separated by raising or lowering heald frames (heddles) to form a clear space where the pick can pass

- Picking: where the weft or pick is propelled across the loom by hand, an air-jet, a rapier or a shuttle.

- Beating-up or battening: where the weft is pushed up against the fell of the cloth by the reed.

The warp is divided into two overlapping groups, or lines (most often adjacent threads belonging to the opposite group) that run in two planes, one above another, so the shuttle can be passed between them in a straight motion. Then, the upper group is lowered by the loom mechanism, and the lower group is raised (shedding), allowing to pass the shuttle in the opposite direction, also in a straight motion. Repeating these actions form a fabric mesh but without beating-up, the final distance between the adjacent wefts would be irregular and far too large.

The secondary motion of the loom are the:

- Let off Motion: where the warp is let off the warp beam at a regulated speed to make the filling even and of the required design

- Take up Motion: Takes up the woven fabric in a regulated manner so that the density of filling is maintained

The tertiary motions of the loom are the stop motions: to stop the loom in the event of a thread break. The two main stop motions are the

- warp stop motion

- weft stop motion

The principal parts of a loom are the frame, the warp-beam or weavers beam, the cloth-roll (apron bar), the heddles, and their mounting, the reed. The warp-beam is a wooden or metal cylinder on the back of the loom on which the warp is delivered. The threads of the warp extend in parallel order from the warp-beam to the front of the loom where they are attached to the cloth-roll. Each thread or group of threads of the warp passes through an opening (eye) in a heddle. The warp threads are separated by the heddles into two or more groups, each controlled and automatically drawn up and down by the motion of the heddles. In the case of small patterns the movement of the heddles is controlled by "cams" which move up the heddles by means of a frame called a harness; in larger patterns the heddles are controlled by a dobby mechanism, where the healds are raised according to pegs inserted into a revolving drum. Where a complex design is required, the healds are raised by harness cords attached to a Jacquard machine. Every time the harness (the heddles) moves up or down, an opening (shed) is made between the threads of warp, through which the pick is inserted. Traditionally the weft thread is inserted by a shuttle.

On a conventional loom, the weft thread is carried on a pirn, in a shuttle that passes through the shed. A handloom weaver could propel the shuttle by throwing it from side to side with the aid of a picking stick. The "picking" on a power loom is done by rapidly hitting the shuttle from each side using an overpick or underpick mechanism controlled by cams 80-250 times a minute. When a pirn is depleted, it is ejected from the shuttle and replaced with the next pirn held in a battery attached to the loom. Multiple shuttle boxes allow more than one shuttle to be used. Each can carry a different colour which allows banding across the loom.

The rapier-type weaving machines do not have shuttles, they propel the weft by means of small grippers or rapiers that pick up the filling thread and carry it halfway across the loom where another rapier picks it up and pulls it the rest of the way. Some carry the filling yarns across the loom at rates in excess of 2,000 metres per minute. Manufacturers such as Picanol have reduced the mechanical adjustments to a minimum, and control all the functions through a computer with a graphical user interface. Other types use compressed air to insert the pick. They are all fast, versatile and quiet.

The warp is sized in a starch mixture for smoother running. The loom warped (loomed or dressed) by passing the sized warp threads through two or more heddles attached to harnesses. The power weavers loom is warped by separate workers. Most looms used for industrial purposes have a machine that ties new warps threads to the waste of previously used warps threads, while still on the loom, then an operator rolls the old and new threads back on the warp beam. The harnesses are controlled by cams, dobbies or a Jacquard head.

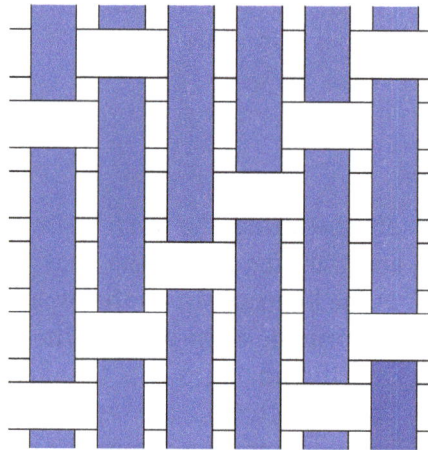

A 3/1 twill, as used in denim

The raising and lowering sequence of warp threads in various sequences gives rise to many possible weave structures: textile production went through profound changes brought about by the industrial revolution in the 19th century. At the beginning of the century in America, weaving was still done by hand, both commercially and at home. Most professional weavers were men who did their work for sale. Women wove items at home for family use. By the end of the 19th century weavers were simply mill workers who tended several water or steam powered looms at a time. The increased speed of production brought more textiles to the average farmhouse and renderd

- plain weave: plain, and hopsacks, poplin, taffeta, poult-de-soie, pibiones and grosgrain.

- twill weave: these are described by weft float followed by warp float, arranged to give diagonal pattern. 2/1 twill, 3/3 twill, 1/2 twill. These are softer fabrics than plain weaves.,

- satin weave: satins and sateens,

- complex computer-generated interlacings.

- pile fabrics : such as velvets and velveteens

Both warp and weft can be visible in the final product. By spacing the warp more closely, it can completely cover the weft that binds it, giving a *warp faced* textile such as repp weave. Conversely, if the warp is spread out, the weft can slide down and completely cover the warp, giving a *weft faced* textile, such as a tapestry or a Kilim rug. There are a variety of loom styles for hand weaving and tapestry.

History

There are some indications that weaving was already known in the Paleolithic era, as early as 27,000 years ago. An indistinct textile impression has been found at the Dolní Věstonice site. According to the find, the weavers of Upper Palaeolithic were manufacturing a variety of cordage types, produced plaited basketry and sophisticated twined and plain woven cloth. The artifacts include imprints in clay and burned remnants of cloth.

The oldest known textiles found in the Americas are remnants of six finely woven textiles and cordage found in Guitarrero Cave, Peru. The weavings, made from plant fibres, are dated between 10100 and 9080 BCE.

Middle East

The earliest known Neolithic textile production in the Old World is supported by a 2013 find of a piece of cloth woven from hemp, in burial F. 7121 at the Çatalhöyük site suggested to be from around 7000 B.C. Further finds come from the advanced civilisation preserved in the pile dwellings in Switzerland. Another extant fragment from the Neolithic was found in Fayum, at a site dated to about 5000 BCE. This fragment is woven at about 12 threads by 9 threads per cm in a plain weave. Flax was the predominant fibre in Egypt at this time (3600 BCE) and continued popularity in the Nile Valley, though wool became the primary fibre used in other cultures around 2000 BCE. Weaving was known in all the great civilisations, but no clear line of causality has been established. Early looms required two people to create the shed and one person to pass through the filling. Early looms wove a fixed length of cloth, but later ones allowed warp to be wound out as the fell progressed. The weavers were often children or slaves. Weaving became simpler when the warp was sized.

The Americas

The Indigenous people of the Americas wove textiles of cotton throughout tropical and subtropical America and in the South American Andes of wool from camelids, primarily domesticated llamas and alpacas. Cotton and the camelids were both domesticated by about 4,000 BCE. American weavers are "credited with independently inventing nearly every non-mechanized technique known today." In the Inca Empire of the Andes, women did most of the weaving using backstrap looms to make small pieces of cloth and vertical frame and single-heddle looms for larger pieces. Andean textile weavings were of practical, symbolic, religious, and ceremonial importance and used as currency, tribute, and as a determinant of social class and rank. Sixteenth-century Spanish colonists were impressed by both the quality and quantity of textiles produced by the Inca Empire. Some of the techniques and designs are still in use in the 21st century.

Example of weaving characteristic of Andean civilizations

Tunic woven for Inca leader.

The oldest-known weavings in North America come from the Windover Archaeological Site in Florida. Dating from 4900 to 6500 B.C. and made from plant fibres, the Windover hunter-gatherers produced "finely crafted" twined and plain weave textiles.

Girls weaving a Persian rug in Hamadan, circa 1922.
Note the design templates (called 'cartoons') at top of loom.

China and East Asia

The weaving of silk from silkworm cocoons has been known in China since about 3500 BCE. Silk that was intricately woven and dyed, showing a well developed craft, has been found in a Chinese tomb dating back to 2700 BCE.

Sericulture and silk weaving spread to Korea by 200 BCE, to Khotan by 50 CE, and to Japan by about 300 CE.

The pit-treadle loom may have originated in India though most authorities establish the invention in China. Pedals were added to operate heddles. By the Middle Ages such devices also appeared in Persia, Sudan, Egypt and possibly the Arabian Peninsula, where "the operator sat with his feet in a pit below a fairly low-slung loom." In 700 CE, horizontal looms and vertical looms could be found in many parts of Asia, Africa and Europe. In Africa, the rich dressed in cotton while the poorer wore wool. By the 12th century it had come to Europe either from the Byzantium or Moorish Spain where the mechanism was raised higher above the ground on a more substantial frame.

Medieval Europe

The predominant fibre was wool, followed by linen and nettlecloth for the lower classes. Cotton was introduced to Sicily and Spain in the 9th century. When Sicily was captured by the Normans, they took the technology to Northern Italy and then the rest of Europe. Silk fabric production was reintroduced towards the end of this period and the more sophisticated silk weaving techniques were applied to the other staples.

The weaver worked at home and marketed his cloth at fairs. Warp-weighted looms were commonplace in Europe before the introduction of horizontal looms in the 10th and 11th centuries. Weaving became an urban craft and to regulate their trade, craftsmen applied to establish a guild. These initially were merchant guilds, but developed into separate trade guilds for each skill. The cloth merchant who was a member of a city's weavers guild was allowed to sell cloth; he acted as a middleman between the tradesmen weavers and the purchaser. The trade guilds controlled quality and the training needed before an artisan could call himself a weaver.

Weaver, Nürnberg, c. 1425

By the 13th century, an organisational change took place, and a system of putting out was introduced. The cloth merchant purchased the wool and provided it to the weaver, who sold his produce back to the merchant. The merchant controlled the rates of pay and economically dominated the cloth industry. The merchants' prosperity is reflected in the wool towns of eastern England; Norwich, Bury St Edmunds and Lavenham being good examples. Wool was a political issue. The supply of thread has always limited the output of a weaver. About that time, the spindle method of spinning was replaced by the great wheel and soon after the treadle-driven spinning wheel. The loom remained the same but with the increased volume of thread it could be operated continuously.

The 14th century saw considerable flux in population. The 13th century had been a period of relative peace; Europe became overpopulated. Poor weather led to a series of poor harvests and starvation. There was great loss of life in the Hundred Years War. Then in 1346, Europe was struck with the Black Death and the population was reduced by up to a half. Arable land was labour-intensive and sufficient workers no longer could be found. Land prices dropped, and land was sold and put to sheep pasture. Traders from Florence and Bruges bought the wool, then sheep-owning landlords started to weave wool outside the jurisdiction of the city and trade guilds. The weavers started by working in their own homes then production was moved into purpose-built buildings. The working hours and the amount of work were regulated. The putting-out system had been replaced by a factory system.

The migration of the Huguenot Weavers, Calvinists fleeing from religious persecution in mainland Europe, to Britain around the time of 1685 challenged the English weavers of cotton, woollen and worsted cloth, who subsequently learned the Huguenots' superior techniques.

Industrial Revolution

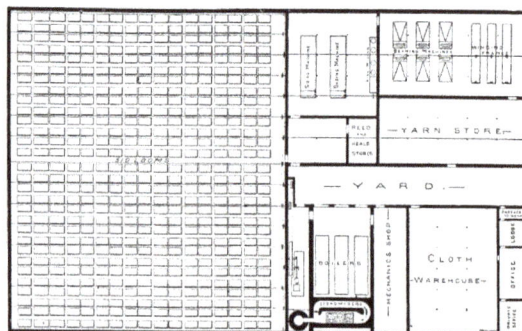

By 1892, most cotton weaving was done in similar weaving sheds, powered by steam.

Before the Industrial Revolution, weaving was a manual craft and wool was the principal staple. In the great wool districts a form of factory system had been introduced but in the uplands weavers worked from home on a putting-out system. The wooden looms of that time might be broad or narrow; broad looms were those too wide for the weaver to pass the shuttle through the shed, so that the weaver needed an expensive assistant (often an apprentice). This ceased to be necessary after John Kay invented the flying shuttle in 1733. The shuttle and the picking stick sped up the process of weaving. There was thus a shortage of thread or a surplus of weaving capacity. The opening of the Bridgewater Canal in June 1761 allowed cotton to be brought into Manchester, an area rich in fast flowing streams that could be used to power machinery. Spinning was the first to be mechanised (spinning jenny, spinning mule), and this led to limitless thread for the weaver.

Edmund Cartwright first proposed building a weaving machine that would function similar to recently developed cotton-spinning mills in 1784, drawing scorn from critics who said the weaving process was too nuanced to automate. He built a factory at Doncaster and obtained a series of patents between 1785 and 1792. In 1788, his brother Major John Cartwight built Revolution Mill at Retford (named for the centenary of the Glorious Revolution). In 1791, he licensed his loom to the Grimshaw brothers of Manchester, but their Knott Mill burnt down the following year (possibly a case of arson). Edmund Cartwight was granted a reward of £10,000 by Parliament for his efforts in 1809. However, success in power-weaving also required improvements by others, including H. Horrocks of Stockport. Only during the two decades after about 1805, did power-weaving take hold. At that time there were 250,000 hand weavers in the UK. Textile manufacture was one of the leading sectors in the British Industrial Revolution, but weaving was a comparatively late sector to be mechanised. The loom became semi-automatic in 1842 with Kenworthy and Bulloughs Lancashire Loom. The various innovations took weaving from a home-based artisan activity (labour-intensive and man-powered) to steam driven factories process. A large metal manufacturing industry grew to produce the looms, firms such as Howard & Bullough of Accrington, and Tweedales and Smalley and Platt Brothers. Most power weaving took place in weaving sheds, in small towns circling Greater Manchester away from the cotton spinning area. The earlier combination mills where spinning and weaving took place in adjacent buildings became rarer. Wool and worsted weaving took place in West Yorkshire and particular Bradford, here there were large factories such as Lister's or Drummond's, where all the processes took place. Both men and women with weaving skills emigrated, and took the knowledge to their new homes in New England, to places like Pawtucket and Lowell.

Woven 'grey cloth' was then sent to the finishers where it was bleached, dyed and printed. Natural dyes were originally used, with synthetic dyes coming in the second half of the 19th century. The need for these chemicals was an important factor in the development of the chemical industry.

The invention in France of the Jacquard loom in about 1803, enabled complicated patterned cloths to be woven, by using punched cards to determine which threads of coloured yarn should appear on the upper side of the cloth. The jacquard allowed individual control of each warp thread, row by row without repeating, so very complex patterns were suddenly feasible. Samples exist showing calligraphy, and woven copies of engravings. Jacquards could be attached to handlooms or powerlooms.

The Role of the Weaver

A distinction can be made between the role and lifestyle and status of a handloom weaver, and that of the powerloom weaver and craft weaver. The perceived threat of the power loom led to disquiet and industrial unrest. Well known protests movements such as the Luddites and the Chartists had hand loom weavers amongst their leaders. In the early 19th century power weaving became viable. Richard Guest in 1823 made a comparison of the productivity of power and hand loom weavers:

> A very good Hand Weaver, a man twenty-five or thirty years of age, will weave two pieces of nine-eighths shirting per week, each twenty-four yards long, and containing one hundred and five shoots of weft in an inch, the reed of the cloth being a forty-four, Bolton count, and the warp and weft forty hanks to the pound, A Steam Loom Weaver, fifteen years of age, will in the same time weave seven similar pieces.

He then speculates about the wider economics of using powerloom weavers:

> ...it may very safely be said, that the work done in a Steam Factory containing two hundred Looms, would, if done by hand Weavers, find employment and support for a population of more than two thousand persons.

Hand Loom Weavers

Hand loom weavers were mainly men- due to the strength needed to batten. They worked from home sometimes in a well lit attic room. The women of the house would spin the thread they needed, and attend to finishing. Later women took to weaving, they obtained their thread from the spinning mill, and working as outworkers on a piecework contract. Over time competition from the power looms drove down the piece rate and they existed in increasing poverty.

Power Loom Weavers

Power loom workers were usually girls and young women. They had the security of fixed hours, and except in times of hardship, such as in the cotton famine, regular income. They were paid a wage and a piece work bonus. Even when working in a combined mill, weavers stuck together and enjoyed a tight-knit community. The women usually minded the four machines and kept the looms oiled and clean. They were assisted by 'little tenters', children on a fixed wage who ran errands and did small tasks. They learnt the job of the weaver by watching. Often they would be half timers, carrying a green card which teacher and overlookers would sign to say they had turned up at the mill in the morning and in the afternoon at the school. At fourteen or so they come full-time into the mill, and started by sharing looms with an experienced worker where it was important to learn quickly as they would both be on piece work. Serious problems with the loom were left to the tackler to sort out. He would inevitably be a man, as were usually the overlookers. The mill had its health and safety issues, there was a reason why the women tied their hair back with scarves. Inhaling cotton dust caused lung problems, and the noise was causing total hearing loss. Weavers would mee-maw as normal conversation was impossible. Weavers used to 'kiss the shutttle' that is suck thread though the eye of the shuttle- this left a foul taste in the mouth due to the oil which was also carcinogenic.

Craft Weavers

Arts and Crafts was an international design philosophy that originated in England and flourished between 1860 and 1910 (especially the second half of that period), continuing its influence until the 1930s. Instigated by the artist and writer William Morris (1834–1896) during the 1860s and inspired by the writings of John Ruskin (1819–1900), it had its earliest and most complete development in the British Isles but spread to Europe and North America. It was largely a reaction against mechanisation and the philosophy advocated of traditional craftsmanship using simple forms and often medieval, romantic or folk styles of decoration. Hand weaving was highly regard and taken up as a decorative art.

Bauhaus Weaving Workshop

In the 1920s the weaving workshop of the Bauhaus design school in Germany aimed to raise weaving, previously seen as a craft, to a fine art, and also to investigate the industrial requirements of modern weaving and fabrics. Under the direction of Gunta Stölzl, the workshop experimented with unorthodox materials, including cellophane, fiberglass, and metal. From expressionist tapestries to the development of soundproofing and light-reflective fabric, the workshop's innovative approach instigated a modernist theory of weaving. Former Bauhaus student and teacher Anni Albers published the seminal 20th-century text *On Weaving* in 1965. Other notables from the Bauhaus weaving workshop include Otti Berger, Margaretha Reichardt, and Benita Otte.

Other Cultures

Weaving in the American Colonies (1500-1800)

Colonial America relied heavily on Great Britain for manufactured goods of all kinds. British policy was to encourage the production of raw materials in colonies and discourage manufacturing. The Wool Act 1699 restricted the export of colonial wool. As a result, many people wove cloth from locally produced fibres. The colonists also used wool, cotton and flax (linen) for weaving, though hemp could be made into serviceable canvas and heavy cloth. They could get one cotton crop each year; until the invention of the cotton gin it was a labour-intensive process to separate the seeds from the fibres.

A plain weave was preferred as the added skill and time required to make more complex weaves kept them from common use. Sometimes designs were woven into the fabric but most were added after weaving using wood block prints or embroidery.

American Southwest

Weaving a traditional Navajo rug

Textile weaving, using cotton dyed with pigments, was a dominant craft among pre-contact tribes of the American southwest, including various Pueblo peoples, the Zuni, and the Ute tribes. The first Spaniards to visit the region wrote about seeing Navajo blankets. With the introduction of Navajo-Churro sheep, the resulting woolen products have become very well known. By the 18th century the Navajo had begun to import yarn with their favorite color, Bayeta red. Using an upright loom, the Navajos wove blankets worn as garments and then rugs after the 1880s for trade. Navajo traded for commercial wool, such as Germantown, imported from Pennsylvania. Under the influence of European-American settlers at trading posts, Navajos created new and distinct styles, including "Two Gray Hills" (predominantly black and white, with traditional patterns), "Teec Nos Pos" (colorful, with very extensive patterns), "Ganado" (founded by Don Lorenzo Hubbell), red dominated patterns with black and white, "Crystal" (founded by J. B. Moore), Oriental and Persian styles (almost always with natural dyes), "Wide Ruins," "Chinlee," banded geometric patterns, "Klagetoh," diamond type patterns, "Red Mesa" and bold diamond patterns. Many of these patterns exhibit a fourfold symmetry, which is thought to embody traditional ideas about harmony, or *hózhó*.

Amazon Cultures

Among the indigenous people of the Amazon basin densely woven palm-bast mosquito netting, or tents, were utilized by the Panoans, Tupinambá, Western Tucano, Yameo, Záparoans, and perhaps by the indigenous peoples of the central Huallaga River basin (Steward 1963:520). Aguaje palm-bast (Mauritia flexuosa, Mauritia minor, or swamp palm) and the frond spears of the Chambira palm (Astrocaryum chambira, A.munbaca, A.tucuma, also known as Cumare or Tucum) have been used for centuries by the Urarina of the Peruvian Amazon to make cordage, net-bags hammocks, and to weave fabric. Among the Urarina, the production of woven palm-fiber goods is imbued with varying degrees of an aesthetic attitude, which draws its authentication from referencing the Urarina's primordial past. Urarina mythology attests to the centrality of weaving and its role in engendering Urarina society. The post-diluvial creation myth accords women's weaving knowledge a pivotal role in Urarina social reproduction. Even though palm-fiber cloth is regularly removed from circulation through mortuary rites, Urarina palm-fiber wealth is neither completely inalienable, nor fungible since it is a fundamental medium for the expression of labor and exchange. The circulation of palm-fiber wealth stabilizes a host of social relationships, ranging from marriage and fictive kinship (*compadrazco*, spiritual compeership) to perpetuating relationships with the deceased.

Loom

A loom is a device used to weave cloth and tapestry. The basic purpose of any loom is to hold the warp threads under tension to facilitate the interweaving of the weft threads. The precise shape of the loom and its mechanics may vary, but the basic function is the same.

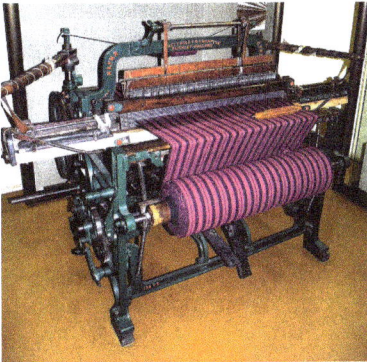

A foot-treadle operated Hattersley & Sons, Domestic Loom, built under license in 1893, in Keighley, Yorkshire.

A woman in Konya, Turkey, works at a vertical loom

Etymology

The word "loom" is derived from the Old English *"geloma"* formed from ge-(perfective prefix) and *"loma"*, a root of unknown origin; this meant utensil or tool or machine of any kind. In 1404 it was used to mean a machine to enable weaving thread into cloth. By 1838 it had gained the meaning of a machine for interlacing thread.

Weaving

Weaving is done by intersecting the longitudinal threads, the warp, i.e. "that which is thrown across", with the transverse threads, the weft, i.e. "that which is woven".

The major components of the loom are the warp beam, heddles, harnesses or shafts (as few as two, four is common, sixteen not unheard of), shuttle, reed and takeup roll. In the loom, yarn processing includes shedding, picking, battening and taking-up operations. These are the principal motions.

- Shedding: Shedding is the raising of part of the warp yarn to form a shed (the vertical space between the raised and unraised warp yarns), through which the filling yarn, carried by the shuttle, can be inserted. On the modern loom, simple and intricate shedding operations are performed automatically by the heddle or heald frame, also known as a harness. This is a rectangular frame to which a series of wires, called heddles or healds, are attached. The yarns are passed through the eye holes of the heddles, which hang vertically from the harnesses. The weave pattern determines which harness controls which warp yarns, and the number of harnesses used depends on the complexity of the weave. Two common methods of controlling the heddles are dobbies and a Jacquard Head.

Shuttles

- Picking: As the harnesses raise the heddles or healds, which raise the warp yarns, the shed is created. The filling yarn is inserted through the shed by a small carrier device called a shuttle. The shuttle is normally pointed at each end to allow passage through the shed. In a traditional shuttle loom, the filling yarn is wound onto a quill, which in turn is mounted in the shuttle. The filling yarn emerges through a hole in the shuttle as it moves across the loom. A single crossing of the shuttle from one side of the loom to the other is known as a pick. As the shuttle moves back and forth across the shed, it weaves an edge, or selvage, on each side of the fabric to prevent the fabric from raveling.

- Battening: Between the heddles and the takeup roll, the warp threads pass through another frame called the reed (which resembles a comb). The portion of the fabric that has already been formed but not yet rolled up on the takeup roll is called the fell. After the shuttle moves across the loom laying down the fill yarn, the weaver uses the reed to press (or batten) each filling yarn against the fell. Conventional shuttle looms can operate at speeds of about 150 to 160 picks per minute.

There are two secondary motions, because with each weaving operation the newly constructed fabric must be wound on a cloth beam. This process is called taking up. At the same time, the warp yarns must be let off or released from the warp beams. To become fully automatic, a loom needs a tertiary motion, the filling stop motion. This will brake the loom, if the weft thread breaks. An automatic loom requires 0.125 hp to 0.5 hp to operate.

Types of Looms

Back Strap Loom

A simple loom which has its roots in ancient civilizations consists of two sticks or bars between which the warps are stretched. One bar is attached to a fixed object, and the other to the weaver usually by means of a strap around the back. On traditional looms, the two main sheds are operated by means of a shed roll over which one set of warps pass, and continuous string heddles which encase each of the warps in the other set. The weaver leans back and uses his or her body weight to tension the loom. To open the shed controlled by the string heddles, the weaver relaxes tension on the warps and raises the heddles. The other shed is usually opened by simply drawing the shed roll toward the weaver. Both simple and complex textiles can be woven on this loom. Width is limited to how far the weaver can reach from side to side to pass the shuttle. Warp faced textiles, often decorated with intricate pick-up patterns woven in complementary and supplementary warp techniques are woven by indigenous peoples today around the world. They produce such things as belts, ponchos, bags, hatbands and carrying cloths. Supplementary weft patterning and brocading is practiced in many regions. Balanced weaves are also possible on the backstrap loom. Today, commercially produced backstrap loom kits often include a rigid heddle.

Warp-weighted Loom

The warp-weighted loom is a vertical loom that may have originated in the Neolithic period. The earliest evidence of warp-weighted looms comes from sites belonging to the Starčevo culture in modern Serbia and Hungary and from late Neolithic sites in Switzerland. This loom was used in Ancient Greece, and spread north and west throughout Europe thereafter. Its defining characteristic is hanging weights (loom weights) which keep bundles of the warp threads taut. Frequently, extra warp thread is wound around the weights. When a weaver has reached the bottom of the available warp, the completed section can be rolled around the top beam, and additional lengths of warp threads can be unwound from the weights to continue. This frees the weaver from vertical size constraints.

Drawloom

A drawloom is a hand-loom for weaving figured cloth. In a drawloom, a "figure harness" is used to control each warp thread separately. A drawloom requires two operators, the weaver and an assistant called a "drawboy" to manage the figure harness.

Handloom

Elements of a foot-treadle floor loom

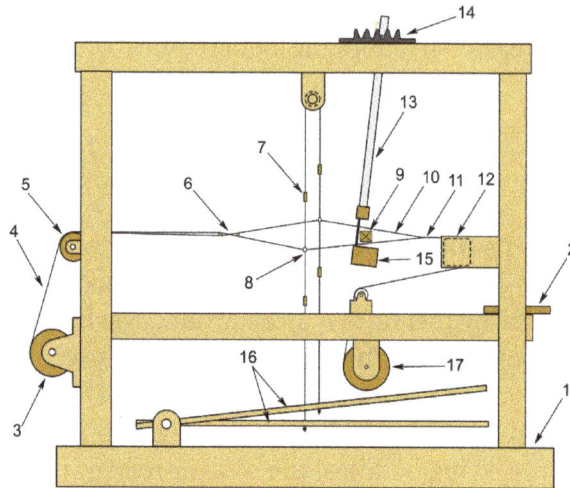

1. Wood frame
2. Seat for weaver
3. Warp beam- let off
4. Warp threads
5. Back beam or platen
6. Rods – used to make a shed
7. Heddle frame - heald frame - harness
8. Heddle- heald - the eye
9. Shuttle with weft yarn
10. Shed
11. Completed fabric
12. Breast beam
13. Batten with reed comb
14. Batten adjustment
15. Lathe
16. Treadles
17. Cloth roll- takeup

A handloom is a simple machine used for weaving. In a wooden vertical-shaft looms, the heddles are fixed in place in the shaft. The warp threads pass alternately through a heddle, and through a space between the heddles (the shed), so that raising the shaft raises half the threads (those passing through the heddles), and lowering the shaft lowers the same threads — the threads passing through the spaces between the heddles remain in place. This was a great discovery in the 13th century.

Flying Shuttle

Hand weavers could only weave a cloth as wide as their armspan. If cloth needed to be wider, two people would do the task (often this would be an adult with a child). John

Kay (1704–1779) patented the flying shuttle in 1733. The weaver held a picking stick that was attached by cords to a device at both ends of the shed. With a flick of the wrist, one cord was pulled and the shuttle was propelled through the shed to the other end with considerable force, speed and efficiency. A flick in the opposite direction and the shuttle was propelled back. A single weaver had control of this motion but the flying shuttle could weave much wider fabric than an arm's length at much greater speeds than had been achieved with the hand thrown shuttle.

The *flying shuttle* was one of the key developments in weaving that helped fuel the Industrial Revolution. The whole picking motion no longer relied on manual skill and it was just a matter of time before it could be powered.

Haute-lisse and Basse-lisse Looms

Looms used for weaving traditional tapestry are classified as *haute-lisse* looms, where the warp is suspended vertically between two rolls, and the *basse-lisse* looms, where the warp extends horizontally between the rolls.

Traditional Looms

Several other types of hand looms exist, including the simple frame loom, pit loom, free-standing loom, and the pegged loom. Each of these can be constructed, and provide work and income in developing societies.

Power Looms

Edmund Cartwright built and patented a power loom in 1785, and it was this that was adopted by the nascent cotton industry in England. The silk loom made by Jacques Vaucanson in 1745 operated on the same principles but was not developed further. The invention of the flying shuttle by John Kay was critical to the development of a commercially successful power loom. Cartwright's loom was impractical but the ideas behind it were developed by numerous inventors in the Manchester area of England where, by 1818, there were 32 factories containing 5,732 looms.

Horrocks loom was viable, but it was the Roberts Loom in 1830 that marked the turning point. Incremental changes to the three motions continued to be made. The problems of sizing, stop-motions, consistent take-up, and a temple to maintain the width remained. In 1841, Kenworthy and Bullough produced the Lancashire Loom which was self-acting or semi-automatic. This enables a youngster to run six looms at the same time. Thus, for simple calicos, the power loom became more economical to run than the hand loom – with complex patterning that used a dobby or Jacquard head, jobs were still put out to handloom weavers until the 1870s. Incremental changes were made such as the Dickinson Loom, culminating in the Keighley-born inventor Northrop, who was working for the Draper Corporation in Hopedale producing the fully automatic Northrop Loom. This loom recharged the shuttle when the pirn was empty. The Draper

E and X models became the leading products from 1909. They were challenged by synthetic fibres such as rayon.

From 1942 the faster and more efficient shuttleless Sulzer looms and the rapier looms were introduced. Modern industrial looms can weave at 2,000 weft insertions per minute.

Weft Insertion

A Picanol rapier loom

Different types of looms are most often defined by the way that the weft, or pick, is inserted into the warp. Many advances in weft insertion have been made in order to make manufactured cloth more cost effective. There are five main types of weft insertion and they are as follows:

- Shuttle: The first-ever powered looms were shuttle-type looms. Spools of weft are unravelled as the shuttle travels across the shed. This is very similar to projectile methods of weaving, except that the weft spool is stored on the shuttle. These looms are considered obsolete in modern industrial fabric manufacturing because they can only reach a maximum of 300 picks per minute.

- Air jet: An air-jet loom uses short quick bursts of compressed air to propel the weft through the shed in order to complete the weave. Air jets are the fastest traditional method of weaving in modern manufacturing and they are able to achieve up to 1,500 picks per minute. However, the amounts of compressed air required to run these looms, as well as the complexity in the way the air jets are positioned, make them more costly than other looms.

- Water jet: Water-jet looms use the same principle as air-jet looms, but they take advantage of pressurized water to propel the weft. The advantage of this type of weaving is that water power is cheaper where water is directly available on site. Picks per minute can reach as high as 1,000.

- Rapier loom: This type of weaving is very versatile, in that rapier looms can weave using a large variety of threads. There are several types of rapiers, but they all use a hook system attached to a rod or metal band to pass the pick across the shed. These machines regularly reach 700 picks per minute in normal production.

- Projectile: Projectile looms utilize an object that is propelled across the shed, usually by spring power, and is guided across the width of the cloth by a series of reeds. The projectile is then removed from the weft fibre and it is returned to the opposite side of the machine so it can get reused. Multiple projectiles are in use in order to increase the pick speed. Maximum speeds on these machines can be as high as 1,050 ppm.

Shedding

Dobby Looms

A dobby loom is a type of floor loom that controls the whole warp threads using a dobby head. Dobby is a corruption of "draw boy" which refers to the weaver's helpers who used to control the warp thread by pulling on draw threads. A dobby loom is an alternative to a treadle loom, where multiple heddles (shafts) were controlled by foot treadles – one for each heddle.

Jacquard Looms

The Jacquard loom is a mechanical loom, invented by Joseph Marie Jacquard in 1801, which simplifies the process of manufacturing textiles with complex patterns such as brocade, damask and matelasse. The loom is controlled by punched cards with punched holes, each row of which corresponds to one row of the design. Multiple rows of holes are punched on each card and the many cards that compose the design of the textile are strung together in order. It is based on earlier inventions by the Frenchmen Basile Bouchon (1725), Jean Baptiste Falcon (1728) and Jacques Vaucanson (1740) To call it a loom is a misnomer, a Jacquard head could be attached to a power loom or a hand loom, the head controlling which warp thread was raised during shedding. Multiple shuttles could be used to control the colour of the weft during picking.

Hand operated Jacquard looms in the Textile Department of the Strzemiński Academy of Fine Arts in Łódź, Poland.

Battening on a jacquard loom in Łódź.

Following the pattern, holes are punched in the appropriate places on a jacquard card.

Circular Looms

A circular loom is used to create a seamless tube of fabric for products such as hosiery, sacks, clothing, fabric hose (such as fire hose) and the like. Circular looms can be small jigs used for Circular knitting or large high-speed machines for modern garments. Modern circular looms use up to ten shuttles driven from below in a circular motion by electromagnets for the weft yarns, and cams to control the warp threads. The warps rise and fall with each shuttle passage, unlike the common practice of lifting all of them at once.

Weaving Technology

Weaving is the most popular way of fabric manufacturing. It is primarily done by interlacing two orthogonal sets (warp and weft) of yarns in a regular and recurring pattern. Actual weaving process is preceded by yarn preparation processes namely winding, warping, sizing, drawing and denting.

Winding converts the smaller ringframe packages to bigger cheeses and cones while removing objectinable yarn faults. Pirn winding is performed to supply the weft yarns in shuttle looms. Figure shows various yarn packages used in textile operations (from left to right: ringframe bobbin or cop, cone, cheese and pirn). Warping is done with the objective to prepare a warper's beam which contains a large number of parallel ends in a double flanged beam. Sizing is the process of applying a protective coating on the warp yarns so that they can withstand repeated stresses, strains and flexing during the weaving process. Finally the fabric is manufactured on looms which perform several operations at proper sequence so that there is interlacement between warp and weft yarns and continuous fabric production.

Types of yarn packages

Inkle Weaving

Inkle weaving is a type of warp-faced weaving where the shed is created by manually raising or lowering the warp yarns, some of which are held in place by fixed heddles on a loom known as an inkle loom. Inkle weaving was referred to in Shakespeare's *Love's*

Labour's Lost. It was brought to the United States in the 1930s, but predates this by many centuries in other countries. The term "Inkle" simply means "ribbon" or "tape" and probably refers to a similarly structured woven good that could have been made on different types of looms, such as a box-loom.

Inkle weaving is commonly used for narrow work such as trims, straps and belts.

Equipment

Inkle looms are constructed in both floor and table-top models. Either model is characterized by a wooden framework upon which dowels have been fastened. These dowels will hold the warp threads when the loom has been dressed.

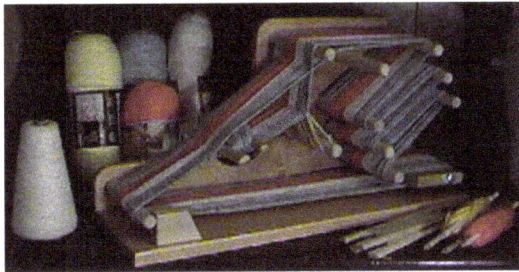

One of the dowels, or a paddle, is constructed so that its position can be adjusted. This tensioning device will be taken in as weaving commences and the warp threads become shorter.

Additional equipment includes yarn of the weaver's choice, yarn or thread for forming heddles and a shuttle to hold the weft. A notebook is also handy for charting weaving diagrams.

Process

The inkle loom is threaded with warp threads according to the weaver's design, alternating between yarn that can be raised and lowered and yarn that is secured in place through the use of the heddles. The raising and lowering of these warp threads creates the shed through which the weft thread will be carried on a shuttle. The weaver should make one pass with the shuttle with each opening of a shed through the raising and lowering of threads.

A simple raising and lowering of threads creates a plain-weave band in which warp threads are slightly offset. Weft threads are only visible at the edges of the band and the weaver may wish to take this into account by warping threads that will form the edges in the same color as the weft.

As the weaving commences, the warp threads will shorten on the loom and the weaver will need to adjust the tension periodically. As the inkle band progresses, it will also get closer to the heddles. The weaver will also need to advance the warp thread along the bottom of the loom to open up new weaving space. In her book "Inkle Weaving," Helene Bress recommends loosening the tension when you are ready to advance the warp. Once you have done so, tighten the tension again and resume your weaving.

There are other more advanced techniques in which, instead of merely allowing warp threads to alternate in their up or down positions, individual threads are brought to the surface to form what is called a "pick up" pattern. One side of the band will show the exposed surfaces of warp threads while, on the other side of the pattern, the weft thread will be visible. Using a supplemental weft thread that will come up over the top of certain warp threads, brocaded designs can also be worked into the inkle band.

An inkle loom is also useful in the practice of tablet weaving for its added portability. Simply thread the warp onto the loom but use cards instead of alternating between free-hanging and heddle-secured yarn.

Uses for Inkle Weaving

Inkle bands are quite strong and can be used in applications where a flat band is desired. Popular modern uses are guitar and camera straps, or, for particularly narrow bands, colorful shoelaces. Traditionally inkle weaving also served as belts and reins. Re-enactors use it as trim for garments and other textiles.

Yarn Count

Yarn count: Yarn count represents the coarseness or fineness of yarns. There are two distinct principles to express the yarn count.

- Direct systems (Example: Tex, Denier)

- Indirect systems (Example: new English i.e. Ne, Metric i.e. Nm)

Direct system revolves around expressing the mass of yarn per unit length. In contrast, indirect system expresses length of yarn per unit mass. For example, 10 tex yarn implies that the 1000 m long yarn will have a mass of 10 g. Similarly, for 10 denier, 9000 m long yarn will have a mass of 10 g. Denier is popularly used to express the fineness of synthetic fibres and filaments. A 10 denier yarn is finer than a 10 tex yarn as for the same mass, the length is nine times for the former.

On the other hand, 10 Ne implies that 1 pound yarn will have a length of 10 × 840 yards. As the Ne value increases (say from 10 Ne to 20 Ne), the yarn becomes finer.

Table: Direct and indirect systems of yarn count

TYPE	NAME	UNIT OF MASS	UNIT OF LENGTH
Direct	Tex	Gram	1000 m
	Denier	Gram	9000 m
Indirect	Ne	Pound	Hank (840 yards)
	Metric	Kilogram	Kilometer

Knitting

Knitting is a method by which yarn is manipulated to create a textile or fabric for use in many types of garments.

Knitting creates multiple loops of yarn, called stitches, in a line or tube. Knitting has multiple active stitches on the needle at one time. Knitted fabric consists of a number of consecutive rows of interlocking loops. As each row progresses, a newly created loop is pulled through one or more loops from the prior row, placed on the gaining needle, and the loops from the prior row are then pulled off the other needle.

Multi-colored knitwork made in stockinette stitch.

Yarn bombing (or knitted graffiti) in Ibarra de Aramayona (Aramaio)

Knitting may be done by hand or by using a machine.

Different types of yarns (fibre type, texture, and twist), needle sizes, and stitch types may be used to achieve knitted fabrics with different properties (color, texture, weight, heat retention, look, water resistance, and/or integrity).

Structure

Courses and Wales

Like weaving, knitting is a technique for producing a two-dimensional fabric made from a one-dimensional yarn or thread. In weaving, threads are always straight, running parallel either lengthwise (warp threads) or crosswise (weft threads). By contrast, the yarn in knitted fabrics follows a meandering path (a *course*), forming symmetric loops (also called bights) symmetrically above and below the mean path of the yarn. These meandering loops can be easily stretched in different directions giving knit fabrics much more elasticity than woven fabrics. Depending on the yarn and knitting pattern, knitted garments can stretch as much as 500%. For this reason, knitting was initially developed for garments that must be elastic or stretch in response to the wearer's motions, such as socks and hosiery. For comparison, woven garments stretch mainly along one or other of a related pair of directions that lie roughly diagonally between the warp and the weft, while contracting in the other direction of the pair (stretching and contracting with the *bias*), and are not very elastic, unless they are woven from stretchable material such as spandex. Knitted garments are often more form-fitting than woven garments, since their elasticity allows them to contour to the body's outline more closely; by contrast, curvature is introduced into most woven garments only with sewn darts, flares, gussets and gores, the seams of which lower the elasticity of the woven fabric still further. Extra curvature can be introduced into knitted garments without seams, as in the heel of a sock; the effect of darts, flares, etc. can be obtained with short rows or by increasing or decreasing the number of stitches. Thread used in weaving is usually much finer than the yarn used in knitting, which can give the knitted fabric more bulk and less drape than a woven fabric.

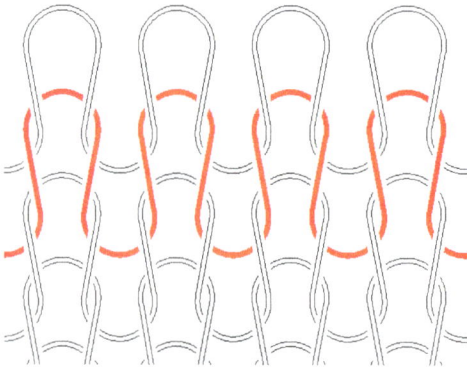

Structure of stockinette, a common knitted fabric. The meandering red path defines one *course*, the path of the yarn through the fabric. The uppermost white loops are unsecured and "active", but they secure the red loops suspended from them. In turn, the red loops secure the white loops just below them, which in turn secure the loops below them, and so on.

Alternating wales of red and yellow knit stitches. Each stitch in a wale is suspended from the one above it.

If they are not secured, the loops of a knitted course will come undone when their yarn is pulled; this is known as *ripping out*, *unravelling* knitting, or humorously, *frogging* (because you 'rip it', this sounds like a frog croaking: 'rib-bit'). To secure a stitch, at least one new loop is passed through it. Although the new stitch is itself unsecured ("active" or "live"), it secures the stitch(es) suspended from it. A sequence of stitches in which each stitch is suspended from the next is called a *wale*. To secure the initial stitches of a knitted fabric, a method for casting on is used; to secure the final stitches in a wale, one uses a method of binding/casting off. During knitting, the active stitches are secured mechanically, either from individual hooks (in knitting machines) or from a knitting needle or frame in hand-knitting.

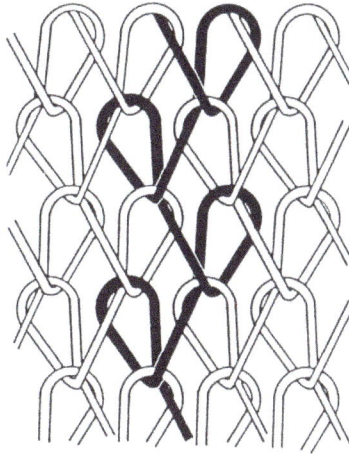

Basic pattern of warp knitting. Parallel yarns zigzag lengthwise along the fabric, each loop securing a loop of an adjacent strand from the previous row.

Weft and Warp Knitting

There are two major varieties of knitting: weft knitting and warp knitting. In the more common *weft knitting*, the wales are perpendicular to the course of the yarn. In warp knitting, the wales and courses run roughly parallel. In weft knitting, the entire fabric may be produced from a single yarn, by adding stitches to each wale in turn, moving across the fabric as in a raster scan. By contrast, in warp knitting, one yarn is required for every wale. Since a typical piece of knitted fabric may have hundreds of wales, warp knitting is typically done by machine, whereas weft knitting is done by both hand and machine. Warp-knitted fabrics such as tricot and milanese are resistant to runs, and are commonly used in lingerie.

Weft-knit fabrics may also be knit with multiple yarns, usually to produce interesting color patterns. The two most common approaches are intarsia and stranded colorwork. In intarsia, the yarns are used in well-segregated regions, e.g., a red apple on a field of green; in that case, the yarns are kept on separate spools and only one is knitted at any time. In the more complex stranded approach, two or more yarns alternate repeatedly within one row and all the yarns must be carried along the row, as seen in Fair

Isle sweaters. Double knitting can produce two separate knitted fabrics simultaneously (e.g., two socks). However, the two fabrics are usually integrated into one, giving it great warmth and excellent drape.

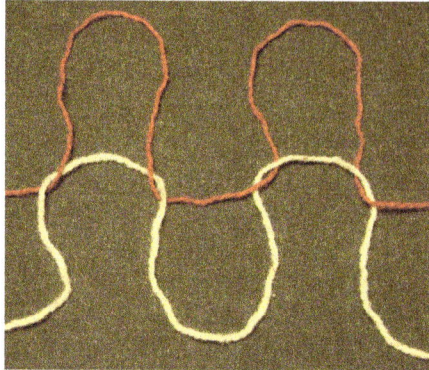

In the knit stitch on the left, the next (red) loop passes through the previous (yellow) loop from *below*, whereas in the purl stitch (right), the next stitch enters from above. Thus, a knit stitch on one side of the fabric appears as a purl stitch on the other, and vice versa.

Knit and Purl Stitches

In securing the previous stitch in a wale, the next stitch can pass through the previous loop from either below or above. If the former, the stitch is denoted as a *knit stitch* or a *plain stitch*; if the latter, as a *purl stitch*. The two stitches are related in that a knit stitch seen from one side of the fabric appears as a purl stitch on the other side.

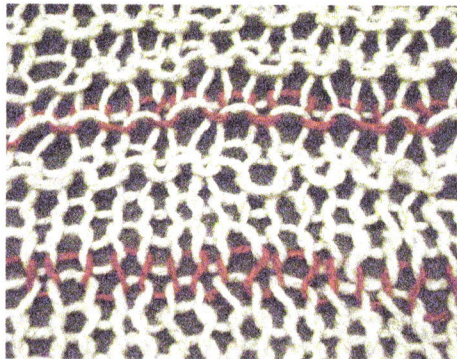

Two courses of red yarn illustrating two basic fabric types. The lower red course is knit into the white row below it and is itself knit on the next row; this produces *stockinette* stitch. The upper red course is purled into the row below and then is knit, consistent with *garter* stitch.

The two types of stitches have a different visual effect; the knit stitches look like "V"'s stacked vertically, whereas the purl stitches look like a wavy horizontal line across the fabric. Patterns and pictures can be created in knitted fabrics by using knit and purl stitches as "pixels"; however, such pixels are usually rectangular, rather than square, depending on the gauge/tension of the knitting. Individual stitches, or rows of stitches, may be made taller by drawing more yarn into the new loop (an elongated stitch), which is the basis for uneven knitting: a row of tall stitches may alternate with one or more rows of short stitches for an interesting visual effect. Short and tall stitches may also alternate within a row, forming a fish-like oval pattern.

A dropped stitch, or missed stitch, is a common error that creates an extra loop to be fixed.

In the simplest knitted fabrics, all the stitches are knit or purl; this is known as a garter stitch. Alternating rows of knit stitches and purl stitches produce what is known as a stockinette pattern/stocking stitch. Vertical stripes (ribbing) are possible by having alternating wales of knit and purl stitches. For example, a common choice is 2x2 ribbing, in which two wales of knit stitches are followed by two wales of purl stitches, etc. Horizontal striping (welting) is also possible, by alternating *rows* of knit and purl stitches. Checkerboard patterns (basketweave) are also possible, the smallest of which is known as *seed/moss stitch*: the stitches alternate between knit and purl in every wale and along every row.

Fabrics in which the number of knit and purl stitches are not the same, such as stockinette/stocking stitch, have a tendency to curl; by contrast, those in which knit and purl stitches are arranged symmetrically (such as ribbing, garter stitch or seed/moss stitch) tend to lie flat and drape well. Wales of purl stitches have a tendency to recede, whereas those of knit stitches tend to come forward. Thus, the purl wales in ribbing tend to be invisible, since the neighboring knit wales come forward. Conversely, rows of purl stitches tend to form an embossed ridge relative to a row of knit stitches. This is the basis of shadow knitting, in which the appearance of a knitted fabric changes when viewed from different directions.

Typically, a new stitch is passed through a single unsecured ("active") loop, thus lengthening that wale by one stitch. However, this need not be so; the new loop may be passed through an already secured stitch lower down on the fabric, or even between secured stitches (a dip stitch). Depending on the distance between where the loop is drawn through the fabric and where it is knitted, dip stitches can produce a subtle stippling or long lines across the surface of the fabric, e.g., the lower leaves of a flower. The new loop may also be passed between two stitches in the *present* row, thus clustering the intervening stitches; this approach is often used to produce a smocking effect in the fabric. The new loop may also be passed through *two or more* previous stitches, producing a decrease and merging wales together. The merged stitches need not be from the same row; for example, a tuck can be formed by knitting stitches together from two different rows, producing a raised horizontal welt on the fabric.

Not every stitch in a row need be knitted; some may be left "as is" and knitted on a subsequent row. This is known as slip-stitch knitting. The slipped stitches are naturally longer than the knitted ones. For example, a stitch slipped for one row before knitting would be roughly twice as tall as its knitted counterparts. This can produce interesting visual effects, although the resulting fabric is more rigid because the slipped stitch "pulls" on its neighbours and is less deformable. Mosaic knitting is a form of slip-stitch knitting that knits alternate colored rows and uses slip stitches to form patterns; mosaic-knit fabrics tend to be stiffer than patterned fabrics produced by other methods such as Fair-Isle knitting.

In some cases, a stitch may be deliberately left unsecured by a new stitch and its wale allowed to disassemble. This is known as drop-stitch knitting, and produces a vertical ladder of see-through holes in the fabric, corresponding to where the wale had been.

Right- and Left-plaited Stitches

The stitches on the right are right-plaited, whereas the stitches on the left are left-plaited.

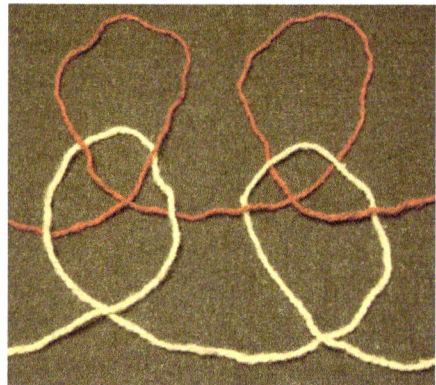

Within limits, an arbitrary number of twists may be added to new stitches, whether they be knit or purl. Here, a single twist is illustrated, with left-plaited and right-plaited stitches on the left and right, respectively.

Both knit and purl stitches may be twisted: usually once if at all, but sometimes twice and (very rarely) thrice. When seen from above, the twist can be clockwise (right yarn over left) or counterclockwise (left yarn over right); these are denoted as right- and left-plaited stitches, respectively. Hand-knitters generally produce right-plaited stitches by knitting or purling through the back loops, i.e., passing the needle through the initial stitch in an unusual way, but wrapping the yarn as usual. By contrast, the left-plaited stitch is generally formed by hand-knitters by wrapping the yarn in the opposite way, rather than by any change in the needle. Although they are mirror images in form, right- and left-plaited stitches are functionally equivalent. Both types of plaited stitches give a subtle but interesting visual texture, and tend to draw the fabric inwards, making it stiffer. Plaited stitches are a common method for knitting jewelry from fine metal wire.

Edges and Joins between Fabrics

The initial and final edges of a knitted fabric are known as the *cast-on* and *bound/cast-off* edges. The side edges are known as the *selvages*; the word derives from "self-edges", meaning that the stitches do not need to be secured by anything else. Many types of selvages have been developed, with different elastic and ornamental properties.

Illustration of entrelac. The blue and white wales are parallel to each other,
but both are perpendicular to the black and gold wales, resembling basket weaving.

Vertical and horizontal edges can be introduced within a knitted fabric, e.g., for button holes, by binding/casting off and re-casting on again (horizontal) or by knitting the fabrics on either side of a vertical edge separately.

Two knitted fabrics can be joined by embroidery-based grafting methods, most commonly the Kitchener stitch. New wales can be begun from any of the edges of a knitted fabric; this is known as picking up stitches and is the basis for entrelac, in which the wales run perpendicular to one another in a checkerboard pattern.

Illustration of cable knitting. The central braid is formed from 2x2 ribbing in which the background is formed of purl stitches and the cables are each two wales of knit stitches. By changing the order in which the stitches are knit, the wales can be made to cross.

Cables, Increases, and Lace

Ordinarily, stitches are knitted in the same order in every row, and the wales of the fabric run parallel and vertically along the fabric. However, this need not be so, since the order in which stitches are knitted may be permuted so that wales cross over one another, forming a cable pattern. Cables patterns tend to draw the fabric together, making it denser and less elastic; Aran sweaters are a common form of knitted cabling. Arbitrarily complex braid patterns can be done in cable knitting, with the proviso that the wales must move ever upwards; it is generally impossible for a wale to move up and then down the fabric. Knitters have developed methods for giving the illusion of a circular wale, such as appear in Celtic knots, but these are inexact approximations. However, such circular wales are possible using Swiss darning, a form of embroidery, or by knitting a tube separately and attaching it to the knitted fabric.

In lace knitting, the pattern is formed by making small, stable holes in the fabric, generally with yarn overs.

A wale can split into two or more wales using increases, most commonly involving a yarn over. Depending on how the increase is done, there is often a hole in the fabric at the point of the increase. This is used to great effect in lace knitting, which consists of making patterns and pictures using such holes, rather than with the stitches themselves. The large and many holes in lacy knitting makes it extremely elastic; for example, some Shetland "wedding-ring" shawls are so fine that they may be drawn through a wedding ring.

By combining increases and decreases, it is possible to make the direction of a wale slant away from vertical, even in weft knitting. This is the basis for bias knitting, and can be used for visual effect, similar to the direction of a brush-stroke in oil painting.

Ornamentations and Additions

Various point-like ornaments may be added to knitting for their look or to improve the wear of the fabric. Examples include various types of bobbles, sequins and beads. Long loops can also be drawn out and secured, forming a "shaggy" texture to the fabric; this is known as loop knitting. Additional patterns can be made on the surface of the knitted fabric using embroidery; if the embroidery resembles knitting, it is often called Swiss darning. Various closures for the garments, such as frogs and buttons can be added; usually buttonholes are knitted into the garment, rather than cut.

Ornamental pieces may also be knitted separately and then attached using applique. For example, differently colored leaves and petals of a flower could be knit separately and attached to form the final picture. Separately knitted tubes can be applied to a knitted fabric to form complex Celtic knots and other patterns that would be difficult to knit.

Unknitted yarns may be worked into knitted fabrics for warmth, as is done in tufting and "weaving" (also known as "couching").

History and Culture

The word is derived from *knot* and ultimately from the Old English *cnyttan*, to knot.

Nålebinding (Danish: literally "binding with a needle" or "needle-binding") is a fabric creation technique predating both knitting and crochet. That said, one of the earliest known examples of true knitting was cotton socks with stranded knit color patterns found in Egypt from the end of the first millennium AD.

The first commercial knitting guilds appear in Western Europe in the early fifteenth century (Tournai in 1429, Barcelona in 1496). The Guild of Saint Fiacre was founded in Paris in 1527 but the archives mention an organization (not necessarily a guild) of knitters from 1268.

With the invention of the knitting machine, knitting "by hand" became a craft used by country people with easy access to fiber. Similar to quilting, spinning, and needlepoint, hand knitting became a leisure activity for the wealthy.

Properties of Fabrics

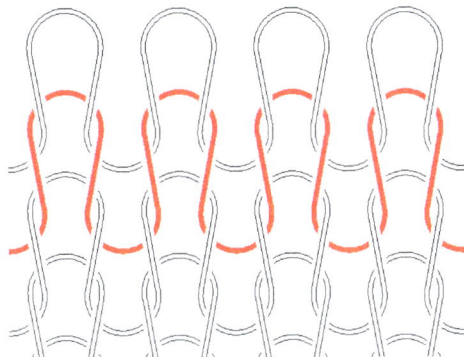

Schematic of stockinette stitch, the most basic weft-knit fabric

The topology of a knitted fabric is relatively complex. Unlike woven fabrics, where strands usually run straight horizontally and vertically, yarn that has been knitted follows a looped path along its row, as with the red strand in the diagram at left, in which the loops of one row have all been pulled through the loops of the row below it.

Because there is no single straight line of yarn anywhere in the pattern, a knitted piece of fabric can stretch in all directions. This elasticity is all but unavailable in woven fabrics which only stretch along the bias. Many modern stretchy garments, even as they rely on elastic synthetic materials for some stretch, also achieve at least some of their stretch through knitted patterns.

Close-up of front of stockinette stitch

Close-up of back of stockinette stitch, also same appearance as reverse stockinette stitch

The basic knitted fabric (as in the diagram, and usually called a *stocking* or *stockinette* pattern) has a definite "right side" and "wrong side". On the right side, the visible portions of the loops are the verticals connecting two rows which are arranged in a grid of *V* shapes. On the wrong side, the ends of the loops are visible, both the tops and bottoms, creating a much more bumpy texture sometimes called *reverse stockinette*. (Despite being the "wrong side," reverse stockinette is frequently used as a pattern in its own right.)

Because the yarn holding rows together is all on the front, and the yarn holding side-by-side stitches together is all on the back, stockinette fabric has a strong tendency to curl toward the front on the top and bottom, and toward the back on the left and right side.

Stitches can be worked from either side, and various patterns are created by mixing regular knit stitches with the "wrong side" stitches, known as purl stitches, either in columns (ribbing), rows (garter, welting), or more complex patterns. Each fabric has different properties: a garter stitch has much more vertical stretch, while ribbing stretches much more horizontally. Because of their front-back symmetry, these two fabrics have little curl, making them popular as edging, even when their stretch properties are not desired.

Different combinations of knit and purl stitches, along with more advanced techniques, generate fabrics of considerably variable consistency, from gauzy to very dense, from highly stretchy to relatively stiff, from flat to tightly curled, and so on.

Close-up of knitting

Texture

The most common texture for a knitted garment is that generated by the flat stockinette stitch as seen, though very small, in machine-made stockings and T-shirts—which is worked in the round as nothing but knit stitches, and worked flat as alternating rows of knit and purl. Other simple textures can be made with nothing but knit and purl stitches, including garter stitch, ribbing, and moss and seed stitches. Adding a "slip stitch" (where a loop is passed from one needle to the other) allows for a wide range of textures, including heel and linen stitches as well as a number of more complicated patterns.

Close-up of ribbing

Some more advanced knitting techniques create a surprising variety of complex textures. Combining certain increases, which can create small eyelet holes in the resulting fabric, with assorted decreases is key to creating knitted lace, a very open fabric resembling lace. Open vertical stripes can be created using the drop-stitch knitting technique. Changing the order of stitches from one row to the next, usually with the help of a cable needle or stitch holder, is key to cable knitting, producing an endless variety of cables, honeycombs, ropes, and Aran sweater patterning. Entrelac forms a rich checkerboard texture by knitting small squares, picking up their side edges, and knitting more squares to continue the piece.

Fair Isle knitting uses two or more colored yarns to create patterns and forms a thicker and less flexible fabric.

The appearance of a garment is also affected by the *weight* of the yarn, which describes the thickness of the spun fibre. The thicker the yarn, the more visible and apparent stitches will be; the thinner the yarn, the finer the texture.

Color

Plenty of finished knitting projects never use more than a single color of yarn, but there are many ways to work in multiple colors. Some yarns are dyed to be either *variegated* (changing color every few stitches in a random fashion) or *self-striping* (changing every few rows). More complicated techniques permit large fields of color (intarsia, for example), busy small-scale patterns of color (such as Fair Isle), or both (double knitting and slip-stitch color, for example).

Yarn with multiple shades of the same hue are called *ombre*, while a yarn with multiple hues may be known as a given *colorway* — a green, red and yellow yarn might be dubbed the "Parrot Colorway" by its manufacturer, for example. *Heathered* yarns contain small amounts of fibre of different colours, while *tweed* yarns may have greater amounts of different colored fibres.

Hand Knitting Process

There are many hundreds of different knitting stitches used by hand knitters. A piece of hand knitting begins with the process of *casting on*, which involves the initial creation of the stitches on the needle. Different methods of casting on are used for different effects: one may be stretchy enough for lace, while another provides a decorative edging. *Provisional* cast-ons are used when the knitting will continue in both directions from the cast-on. There are various methods employed to cast on, such as the "thumb method" (also known as "slingshot" or "long-tail" cast-ons), where the stitches are created by a series of loops that will, when knitted, give a very loose edge ideal for "picking up stitches" and knitting a border; the "double needle method" (also known as "knit-on" or "cable cast-on"), whereby each loop placed on the needle is then "knitted on," which produces a firmer edge ideal on its own as a border; and many more. The number of active stitches remains the same as when cast on unless stitches are added (an increase) or removed (a decrease).

A woman in the process of hand knitting (1904)

Most Western-style hand knitters follow either the English style (in which the yarn is held in the right hand) or the Continental style (in which the yarn is held in the left hand).

There are also different ways to insert the needle into the stitch. Knitting through the front of a stitch is called Western knitting. Going through the back of a stitch is called Eastern knitting. A third method, called combination knitting, goes through the front of a knit stitch and the back of a purl stitch.

Once the hand knitted piece is finished, the remaining live stitches are "cast off". Casting (or "binding") off loops the stitches across each other so they can be removed from the needle without unravelling the item. Although the mechanics are different from casting on, there is a similar variety of methods.

In hand knitting certain articles of clothing, especially larger ones like sweaters, the final knitted garment will be made of several knitted pieces, with individual sections of the garment hand knitted separately and then sewn together. Seamless knitting, where a whole garment is hand knit as a single piece, is also possible. Elizabeth Zimmermann is probably the best-known proponent of seamless or circular hand knitting techniques. Smaller items, such as socks and hats, are usually knit in one piece on double-pointed needles or circular needles. Hats in particular can be started "top down" on double pointed needles with the increases added until the preferred size is achieved, switching to an appropriate circular needle when enough stitches have been added. Care must be taken to bind off at a tension that will allow the "give" needed to comfortably fit on the head.

Mega Knitting

Mega knitting is a term recently coined and relates to the use of knitting needles greater than or equal to half an inch in diameter.

Mega knitting uses the same stitches and techniques as conventional knitting, except that hooks are carved into the ends of the needles. The hooked needles greatly enhance control of the work, catching the stitches and preventing them from slipping off.

It was the development of the knitting machine that introduced hooked needles and enabled faultless, automated knitting. Most knitters probably aren't even aware of the many processes that their fingers perform in the making of a single stitch. However, large gauge needles emphasize those actions and knitting becomes increasingly more awkward when the needle diameter is greater than the width of the knitters finger. On a one-inch diameter (size 50) needle for instance, the shaft begins to taper one and three quarter inches from the tip. This means that the stitches are spread much further apart on mega knitting needles, making them more difficult to control. The hook catches the loop of yarn as each stitch is knitted, meaning that wrists and fingers don't have to work so hard and there is less chance of stitches slipping off the needle. The position of the hook is most important. Turn the left (non-working) hook to face away at all times; turn the right (working) hook toward you up whilst knitting (plain stitch) and away whilst purling.

Mega knitting produces a chunky, bulky fabric or an open lacy weave, depending on the weight and type of yarn used.

Materials

Yarn

Yarn for hand-knitting is usually sold as balls or skeins (hanks), although it may also be wound on spools or cones. Skeins and balls are generally sold with a *yarn-band*, a label that describes the yarn's weight, length, dye lot, fiber content, washing instructions, suggested needle size, likely gauge/tension, etc. It is common practice to save the yarn band for future reference, especially if additional skeins must be purchased. Knitters generally ensure that the yarn for a project comes from a single dye lot. The dye lot specifies a group of skeins that were dyed together and thus have precisely the same color; skeins from different dye-lots, even if very similar in color, are usually slightly different and may produce a visible horizontal stripe when knitted together. If a knitter buys insufficient yarn of a single dye lot to complete a project, additional skeins of the same dye lot can sometimes be obtained from other yarn stores or online. Otherwise, knitters can alternate skeins every few rows to help the dye lots blend together easier.

A hank of wool yarn (center) is uncoiled into its basic loop. A tie is visible at the left; after untying, the hank may be wound into a ball or balls suitable for knitting. Knitting from a normal hank directly is likely to tangle the yarn, producing snarls.

The thickness or weight of the yarn is a significant factor in determining the gauge/ tension, i.e., how many stitches and rows are required to cover a given area for a given stitch pattern. Thicker yarns generally require thicker knitting needles, whereas thinner yarns may be knit with thick or thin needles. Hence, thicker yarns generally require fewer stitches, and therefore less time, to knit up a given garment. Patterns and motifs are coarser with thicker yarns; thicker yarns produce bold visual effects, whereas thinner yarns are best for refined patterns. Yarns are grouped by thickness into six categories: superfine, fine, light, medium, bulky and superbulky; quantitatively, thickness is measured by the number of wraps per inch (WPI). In the British Commonwealth (outside North America) yarns are measured as 1ply, 2ply, 3ply, 4ply, 5ply, 8ply (or double knit),10ply and 12ply (triple knit). The related *weight per unit length* is usually measured in tex or denier.

Transformation of a hank of lavender silk yarn (top) into a ball in which the knitting yarn emerges from the center (bottom). The latter is better for knitting, since the yarn is much less likely to tangle.

Before knitting, the knitter will typically transform a hank/skein into a ball where the yarn emerges from the center of the ball; this making the knitting easier by preventing the yarn from becoming easily tangled. This transformation may be done by hand, or with a device known as a ballwinder. When knitting, some knitters enclose their balls in jars to keep them clean and untangled with other yarns; the free yarn passes through a small hole in the jar-lid.

A yarn's usefulness for a knitting project is judged by several factors, such as its *loft* (its ability to trap air), its *resilience* (elasticity under tension), its washability and colorfastness, its *hand* (its feel, particularly softness vs. scratchiness), its durability against abrasion, its resistance to pilling, its *hairiness* (fuzziness), its tendency to twist or untwist, its overall weight and drape, its blocking and felting qualities, its comfort (breathability, moisture absorption, wicking properties) and of course its look, which includes its color, sheen, smoothness and ornamental features. Other factors include allergenicity; speed of drying; resistance to chemicals, moths, and mildew; melting point and flammability; retention of static electricity; and the propensity to become stained and to accept dyes. Different factors may be more significant than others for different knitting projects, so there is no one "best" yarn. The resilience and propensity to (un)twist are general properties that affect the ease of hand-knitting. More resilient yarns are more forgiving of irregularities in tension; highly twisted yarns are sometimes difficult to knit, whereas untwisting yarns can lead to split stitches, in which not all the yarn is knitted into a stitch. A key factor in knitting is *stitch definition*, corresponding to how well complicated stitch patterns can be seen when made from a given yarn. Smooth, highly spun yarns are best for showing off stitch patterns; at the other extreme, very fuzzy yarns or eyelash yarns have poor stitch definition, and any complicated stitch pattern would be invisible.

The two possible twists of yarn

Although knitting may be done with ribbons, metal wire or more exotic filaments, most yarns are made by spinning fibers. In spinning, the fibers are twisted so that the yarn resists breaking under tension; the twisting may be done in either direction, resulting in a Z-twist or S-twist yarn. If the fibers are first aligned by combing them, the yarn is smoother and called a *worsted*; by contrast, if the fibers are carded but not combed, the yarn is fuzzier and called *woolen-spun*. The fibers making up a yarn may be continuous *filament* fibers such as silk and many synthetics, or they may be *staples* (fibers of an average length, typically a few inches); naturally filament fibers are sometimes cut up into staples before spinning. The strength of the spun yarn against breaking is determined by the amount of twist, the length of the fibers and the thickness of the yarn. In general, yarns become stronger with more twist (also called *worst*), longer fibers and thicker yarns (more fibers); for example, thinner yarns require more twist than do thicker yarns to resist breaking under tension. The thickness of the yarn may vary along its length; a *slub* is a much thicker section in which a mass of fibers is incorporated into the yarn.

The spun fibers are generally divided into animal fibers, plant and synthetic fibers. These fiber types are chemically different, corresponding to proteins, carbohydrates and synthetic polymers, respectively. Animal fibers include silk, but generally are long hairs of animals such as sheep (wool), goat (angora, or cashmere goat), rabbit (angora), llama, alpaca, dog, cat, camel, yak, and muskox (qiviut). Plants used for fibers include cotton, flax (for linen), bamboo, ramie, hemp, jute, nettle, raffia, yucca, coconut husk, banana fiber, soy and corn. Rayon and acetate fibers are also produced from cellulose mainly derived from trees. Common synthetic fibers include acrylics, polyesters such as dacron and ingeo, nylon and other polyamides, and olefins such as polypropylene. Of these types, wool is generally favored for knitting, chiefly owing to its superior elasticity, warmth and (sometimes) felting; however, wool is generally less convenient to clean and some people are allergic to it. It is also common to blend different fibers in the yarn, e.g., 85% alpaca and 15% silk. Even within a type of fiber, there can be great variety in the length and thickness of the fibers; for example, Merino wool and Egyptian cotton are favored because they produce exceptionally long, thin (fine) fibers for their type.

A single spun yarn may be knitted as is, or braided or plied with another. In plying, two or more yarns are spun together, almost always in the opposite sense from which they were spun individually; for example, two Z-twist yarns are usually plied with an S-twist. The opposing twist relieves some of the yarns' tendency to curl up and produces a thicker, *balanced* yarn. Plied yarns may themselves be plied together, producing *cabled yarns* or *multi-stranded yarns*. Sometimes, the yarns being plied are fed at different rates, so that one yarn loops around the other, as in bouclé. The single yarns may be dyed separately before plying, or afterwards to give the yarn a uniform look.

The dyeing of yarns is a complex art that has a long history. However, yarns need not be dyed. They may be dyed just one color, or a great variety of colors. Dyeing may be done industrially, by hand or even hand-painted onto the yarn. A great variety of synthetic dyes have been developed since the synthesis of indigo dye in the mid-19th century; however, natural dyes are also possible, although they are generally less brilliant. The color-scheme of a yarn is sometimes called its colorway. Variegated yarns can produce interesting visual effects, such as diagonal stripes; conversely, a variegated yarn may frustrate an otherwise good knitting pattern by producing distasteful color combination.

Glass/Wax

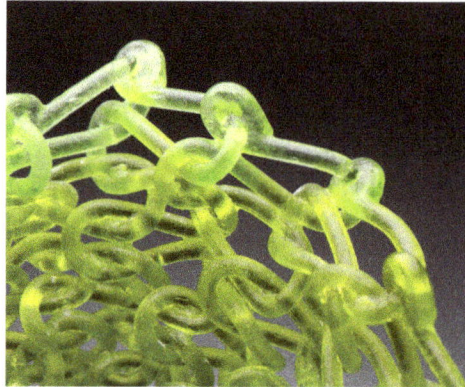

Close-up of 'Jitterbug' - Knitted Glass

Knitted Glass combines knitting, lost-wax casting, mold-making, and kiln-casting. The process involves:

1. *knitting* with wax strands;

2. surrounding the knitted wax piece with a heat-tolerant refractory material;

3. removing the wax by melting it out, thus creating a mold;

4. placing the mold in a kiln where lead crystal glass melts into the mold;

5. after the mold cools, the mold material is removed to reveal the knitted glass piece.

Tools

The process of knitting has three basic tasks:

1. the active (unsecured) stitches must be held so they don't drop

2. these stitches must be released sometime after they are secured

3. new bights of yarn must be passed through the fabric, usually through active stitches, thus securing them.

In very simple cases, knitting can be done without tools, using only the fingers to do these tasks; however, knitting is usually carried out using tools such as knitting needles, knitting machines or rigid frames. Depending on their size and shape, the rigid frames are called stocking frames, knitting boards, knitting rings (also called knitting looms) or knitting spools (also known as knitting knobbies, knitting nancies, or corkers). There is also a technique called knooking of knitting with a crochet hook that has a cord attached to the end, to hold the stitches while they're being worked. Other tools are used to prepare yarn for knitting, to measure and design knitted garments, or to make knitting easier or more comfortable.

Needles

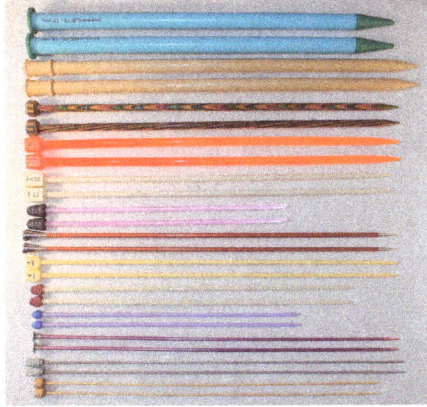

Knitting needles in a variety of sizes and materials. Different materials have varying amounts of friction, and are suitable for different yarn types.

There are three basic types of knitting needles (also called "knitting pins"). The first and most common type consists of two slender, straight sticks tapered to a point at one end, and with a knob at the other end to prevent stitches from slipping off. Such needles are usually 10–16 inches (250–410 mm) long but, due to the compressibility of knitted fabrics, may be used to knit pieces significantly wider. The most important property of needles is their diameter, which ranges from below 2 to 25 mm (roughly 1 inch). The diameter affects the size of stitches, which affects the gauge/tension of the knitting and the elasticity of the fabric. Thus, a simple way to change gauge/tension is to use different needles, which is the basis of uneven knitting. Although the diameter of the knitting needle is often measured in millimeters, there are several measurement systems, particularly those specific to the United States, the United Kingdom and Japan; a conversion table is given at knitting needle. Such knitting needles may be made out of any materials, but the most common materials are metals, wood, bamboo, and plastic. Different materials have different frictions and grip the yarn differently; slick needles such as metallic needles are useful for swift knitting, whereas rougher needles such as bamboo offer more friction and are therefore less prone to dropping stitches. The knitting of new stitches occurs only at the tapered ends. Needles with lighted tips have been sold to allow knitters to knit in the dark.

Double-pointed knitting needles in various materials and sizes. They come in sets of four, five or six.

The second type of knitting needles are straight, double-pointed knitting needles (also called "DPNs"). Double-pointed needles are tapered at both ends, which allows them to be knit from either end. DPNs are typically used for circular knitting, especially smaller tube-shaped pieces such as sleeves, collars, and socks; usually one needle is active while the others hold the remaining active stitches. DPNs are somewhat shorter (typically 7 inches) and are usually sold in sets of four or five.

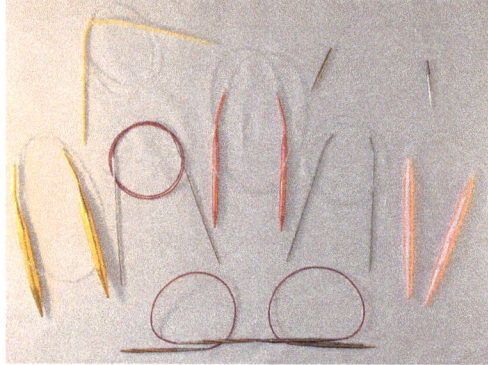

Circular knitting needles in different lengths, materials and sizes,
including plastic, aluminum, steel and nickel-plated brass

Cable needles are a special case of DPNs, although they are usually not straight, but dimpled in the middle. Often, they have the form of a hook. When cabling a knitted piece, a hook is easier to grab and hold the yarn. Cable needles are typically very short (a few inches), and are used to hold stitches temporarily while others are being knitted. Cable patterns are made by permuting the order of stitches; although one or two stitches may be held by hand or knit out of order, cables of three or more generally require a cable needle.

The third needle type consists of circular needles, which are long, flexible double-pointed needles. The two tapered ends (typically 5 inches (130 mm) long) are rigid and straight, allowing for easy knitting; however, the two ends are connected by a flexible strand (usually nylon) that allows the two ends to be brought together. Circular needles are typically 24-60 inches long, and are usually used singly or in pairs; again, the width of the knitted piece may be significantly longer than the length of the circular needle.

A developing trend in the knitting world is interchangeable needles. These kits consist of pairs of needles with usually nylon cables or cords. The cables/cords are screwed into the needles, allowing the knitter to have both flexible straight needles or circular needles. This also allows the knitter to change the diameter and length of the needles as needed.

The ability to work from either end of one needle is convenient in several types of knitting, such as slip-stitch versions of double knitting. Circular needles may be used for flat or circular knitting.

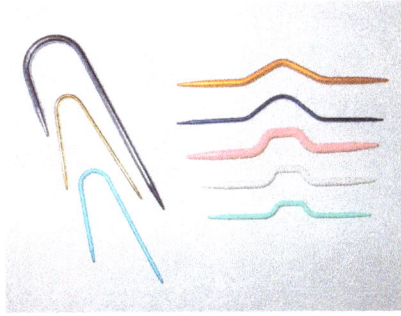
Cable needles

Cable needles are a specific design, and are used to create the twisting motif of a knitted cable. They are made in different sizes, which produces cables of different widths. When in use, the cable needle is used at the same time as two regular needles. It functions by holding together the stitches creating the cable as the other needles create the rest of the stitches for the knitted piece. At specific points indicated by the knitting pattern, the cable needle is moved, the stitches on it are worked by the other needles, then the cable needle is turned around to a different position to create the cable twist.

Mega Needles

Mega knitting needles are generally considered to be any knitting needles larger than size 17 (half inch diameter). Mega needles may or may not have hooks carved in the ends. Hooks on large diameter needles help enormously to control the stitches whilst knitting.

Largest Circular Knitting Needles

The largest aluminum circular knitting needles on record are size US 150 and are nearly 7 feet tall. They are owned by Paradise Fibers and are currently on display in the Paradise Fibers retail showroom.

Record

Julia Hopson with world-record 3.5 meter long knitting needles

The current holder of the Guinness World Record for Knitting with the Largest Knitting Needles is Julia Hopson of Penzance in Cornwall.

Julia knitted a square of ten stitches and ten rows in stockinette stitch using knitting needles that were 6.5 centimeters in diameter and 3.5 meters long.

Ancillary Tools

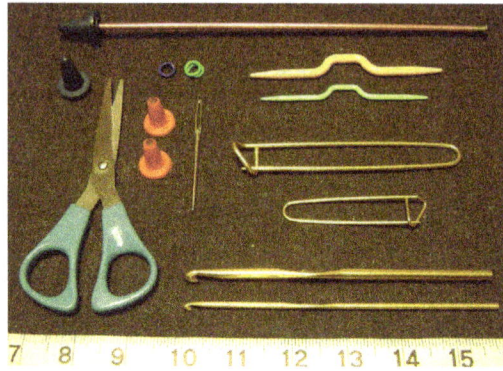

Some ancillary tools used by hand-knitters. Starting from the bottom right are two crochet hooks, two stitch holders (like big blunt safety pins), and two cable needles in pink and green. On the left are a pair of scissors, a yarn needle, green and blue stitch markers, and two orange point protectors. At the top left are two blue point protectors, one on a red needle.

Various tools have been developed to make hand-knitting easier. Tools for measuring needle diameter and yarn properties have been discussed above, as well as the yarn swift, ballwinder and "yarntainers". Crochet hooks and a darning needle are often useful in binding/casting off or in joining two knitted pieces edge-to-edge. The darning needle is used in duplicate stitch (also known as Swiss darning). The crochet hook is also essential for repairing dropped stitches and some specialty stitches such as tufting. Other tools such as knitting spools or pom-pom makers are used to prepare specific ornaments. For large or complex knitting patterns, it is sometimes difficult to keep track of which stitch should be knit in a particular way; therefore, several tools have been developed to identify the number of a particular row or stitch, including circular stitch markers, hanging markers, extra yarn and row counters. A second potential difficulty is that the knitted piece will slide off the tapered end of the needles when unattended; this is prevented by "point protectors" that cap the tapered ends. Another problem is that too much knitting may lead to hand and wrist troubles; for this, special stress-relieving gloves are available. In traditional Shetland knitting a special belt is often used to support the end of one needle allowing the knitting greater speed. Finally, there are sundry bags and containers for holding knitting, yarns and needles.

Commercial Applications

Industrially, metal wire is also knitted into a metal fabric for a wide range of uses including the filter material in cafetieres, catalytic converters for cars and many other uses. These fabrics are usually manufactured on circular knitting machines that would be recognized by conventional knitters as sock machines.

Many fashion designers make heavy use of knitted fabric in their fashion collections. Gordana Gelhausen, who appeared in season six of the television show *Project Runway*, is primarily a knit designer. Other designers and labels that make heavy use of knitting include Michael Kors, Fendi, and Marc Jacobs.

For individual hobbyists, websites such as Etsy, Big Cartel and Ravelry have made it easy to sell knitting patterns on a small scale, in a way similar to eBay.

Graffiti

In the last decade, a practice called knitting graffiti, guerilla knitting, or yarn bombing— the use of knitted or crocheted cloth to modify and beautify one's (usually outdoor) surroundings—emerged in the U.S. and spread worldwide. Magda Sayeg is credited with starting the movement in the US and Knit the City are a prominent group of graffiti knitters in the United Kingdom. Yarn bombers sometimes target existing pieces of graffiti for beautification. For instance, Dave Cole is a contemporary sculpture artist who practiced knitting as graffiti for a large-scale public art installation in Melbourne, Australia for the Big West Arts Festival in 2009. The work was vandalized the night of its completion. A new movie, shot by a Tasmanian filmmaker on a set made almost entirely out of yarn, was partially inspired by "knitted graffiti".

Yarn Crawl

Many major metropolitan cities across the US and Europe host annual Yarn Crawls. The event is typically a multi-day event that caters to all knitters, crochet and yarn enthusiasts that supports the local crafting community. Over the multi-day period, multiple local yarn and knit shops participate in the yarn crawl and offer up store discounts, give away free exclusive patterns, provide classes, trunk shows and conduct raffles for prizes. Participants of the crawl receive a passport and get their passport stamped at each store they visit along the crawl. Traditionally those that get their passports fully stamped are eligible to win a larger gift basket filled with yarn, knitting and crochet goodies. Some local crawls also provide a Knit-Along (KAL) or Crochet-Along (CAL) where attendees follow a specific pattern prior to the crawl and then proudly wear it during the crawl for others to see.

Charity

Hand knitting garments for free distribution to others has become common practice among hand knitting groups. Girls and women hand knitted socks, sweaters, scarves, mittens, gloves, and hats for soldiers in Crimea, the American Civil War, and the Boer Wars; this practice continued in World War I, World War II and the Korean War, and continues for soldiers in Iraq and Afghanistan. The Australian charity *Wrap with Love* continues to provide blankets hand knitted by volunteers to people most in need around the world who have been affected by war.

In the historical projects, yarn companies provided knitting patterns approved by the various branches of the armed services; often they were distributed by local chapters of the American Red Cross. Modern projects usually entail the hand knitting of hats or helmet liners; the liners provided for soldiers must be of 100% worsted weight wool and be crafted using specific colors.

Some charities teach women to knit as a means of clothing their families or supporting themselves.

Clothing and afghans are frequently made for children, the elderly, and the economically disadvantaged in various countries. Pine Ridge Indian Reservation accepts donations for the Lakota people in the United States. Prayer shawls, or shawls in which the crafter meditates or says prayers of their faith while hand knitting with the intent on comforting the recipient, are donated to those experiencing loss or stress. Many knitters today hand knit and donate "chemo caps," soft caps for cancer patients who lose their hair during chemotherapy. Yarn companies offer free knitting patterns for these caps.

Penguin sweaters were hand knitted by volunteers for the rehabilitation of penguins contaminated by exposure to oil slicks. The project is now complete.

Chicken sweaters were also hand knitted to aid battery hens that had lost their feathers. The organization is not currently accepting donations, but maintains a list of volunteers.

Originally started after the 2004 Indonesian tsunami, Knitters Without Borders is a charity challenge issued by knitting personality Stephanie Pearl-McPhee that encourages hand knitters to donate to Médecins Sans Frontières (Doctors Without Borders). Instead of[hand knitting for charity, knitters are encouraged to donate a week's worth of disposable income, including money that otherwise might have been spent on yarn. Knitted items are occasional offered as prizes to donors. As of September 2011, Knitters Without Borders donors have contributed CAD$1,062,217.

Security blankets can also be made through the Project Linus organization which helps needy children.

There are organizations that help reach other countries in need such as afghans for Afghans. This outreach is described as, "afghans for Afghans is a humanitarian and educational people-to-people project that sends hand-knit and crocheted blankets and sweaters, vests, hats, mittens, and socks to the beleaguered people of Afghanistan."

Health Benefits

Studies have shown that hand knitting, along with other forms of needlework, provide several significant health benefits. These studies have found the rhythmic and repetitive action of hand knitting can "help prevent and manage stress, pain and depression, which in turn strengthens the body's immune system", as well as create a relaxation response in the body which can decrease blood pressure, heart rate, help prevent illness, and have a calming effect. Pain specialists have also found that the brain chemistry is changed when one hand knits, resulting in an increase in "feel good" hormones (i.e. serotonin and dopamine), and a decrease in stress hormones.

Hand knitting, along with other leisure activities, has been linked to reducing the risk of developing Alzheimer's disease and dementia. Much like physical activity strengthens the body, mental exercise makes the human brain more resilient.

A repository of research into the effect on health of hand knitting can be found at Stitch links, an organization founded in Bath, England.

The earliest image of circular knitting, from the 15th century AD Buxtehude altarpiece

Knitting also helps in the area of social interaction; knitting provides people with opportunities to socialize with others. Some ways to increase social interaction with knitting is inviting friends over to knit and chat with each other. Even if they've never knitted before this can be a fun way to interact with your friends.

Another interesting way that knitting can positively impact your life is improving the dexterity in your hands and figures. This keeps your fingers limber and can be especially helpful for those with arthritis. Knitting can reduce the pain of arthritis if people make it a daily habit.

Hand Knitting

Hand knitting is a form of knitting, in which the knitted fabric is produced by hand using needles.

Types

Flat Knitting

Flat knitting uses two straight needles to make generally two-dimensional (flat) pieces. Flat knitting is usually used to knit flat pieces like scarves, blankets, afghans, and the backs, fronts and arms of sweaters (pullovers).

Circular knitting on a circular needle

Flat knitting. The loops on the metal needle are the active stitches, and the yarn coming out of the knitting on the right is the working yarn.

In flat knitting, generally stockinette stitch, the hand-knitter knits from right-to-left on one side of the fabric, turns the work (over), and then purls right-to-left back to the starting position. Usually the smooth side of the fabric is considered the *right side*, the one facing outwards for viewing; and the side that faces inwards, towards the body, the ridged side, is known as the *wrong side*. Thus, flat knitting involves knitting each row on the right side, then purling each row on the wrong side, etc. If each row is knit (no purls) this creates garter stitch, which has the same appearance on both sides and creates horizontal ridges offset by valleys, rather than a knit and purl side. Patterned stitching, such as cables, can be accomplished with either flat knitting, or in the round, however the technique must follow the desired pattern.

Circular Knitting

Circular knitting (also called "knitting in the round") creates a seamless tube. Knitting is worked in rounds (the equivalent of rows in flat knitting).

Originally, circular knitting was done using a set of four or five double-pointed knitting needles. Circular needles were later invented making this type of knitting easier. A circular needle resembles two short knitting needles connected by a cable of varying length between them. A circular knitting needle with a long cable can be used in place of straight needles to create larger flat-knitted pieces of fabric. Both types of circular knitting are used in creating pieces that are circular or tube-shaped, such as hats, socks, mittens, sleeves, and entire sweaters.

In circular knitting, the hand-knitter generally knits everything from one side, usually the right side. Circular knitting is usually carried out on a single circular needle. In such cases, the knitter can resort to a variety of alternative techniques, such as double-pointed needles, knitting on two circular needles, a Möbius strip-like "magic needle" approach (commonly known as "Magic Loop"), or careful use of slip-stitch knitting or equivalently double knitting to knit the back and front of the tube.

Fabric Finishing

Felting

Felting is the hand-knitters' term for fulling, a technique for joining knitted or woven animal-yarn fibres. The finished product is put in hot water and agitated until it starts to shrink. The end result typically has a felt-like appearance but has reduced dimensions. Bags, mittens, vests, socks, slippers and hats are just a few items that can be felted.

Needle Felting

Needle felting is a technique used to add decoration to a knitted or felted piece, where raw roving is applied using a very sharp barbed felting needle by repeatedly piercing the roving and background together. Once washed in hot water, the appliqued decoration is fused with the background. Felted knitting can be cut with scissors without concern about fraying.

Process

There are many hundreds of different knitting stitches used by knitters. A piece of knitting begins with the process of *casting on*, which involves the initial creation of the stitches on the needle. Different methods of casting on are used for different effects: one may be stretchy enough for lace, while another provides a decorative edging. *Provisional* cast-ons are used when the knitting will continue in both directions from the cast-on. There are various methods employed to cast on, such as the "thumb method" (also known as "slingshot" or "long-tail" cast-ons), where the stitches are created by a series of loops that will, when knitted, give a very loose edge ideal for "picking up stitches" and knitting a border; the "double needle method" (also known as "knit-on" or "cable cast-on"), whereby each loop placed on the needle is then "knitted on," which produces a firmer edge ideal on its own as a border; and many more. The number of active stitches remains the same as when cast on unless stitches are added (an increase) or removed (a decrease).

Most Western-style knitters follow either the English style (in which the yarn is held in the right hand) or the Continental style (in which the yarn is held in the left hand).

There are also different ways to insert the needle into the stitch. Knitting through the front of a stitch is called Western knitting. Going through the back of a stitch is called Eastern knitting. A third method, called combination knitting, goes through the front of a knit stitch and the back of a purl stitch.

Once the knitted piece is finished, the remaining live stitches are "cast off". Casting (or "binding") off loops the stitches across each other so they can be removed from the needle without unravelling the item. Although the mechanics are different from casting on, there is a similar variety of methods.

In knitting certain articles of clothing, especially larger ones like sweaters, the final knitted garment will be made of several knitted pieces, with individual sections of the garment knit separately and then sewn together. Seamless knitting, where a whole garment is knit as a single piece, is also possible. Elizabeth Zimmermann is probably the best-known proponent of seamless or circular knitting techniques. Smaller items, such as socks and hats, are usually knit in one piece on double-pointed needles or circular needles.

Stitches

There are well-nigh an infinite number of possible combinations of knitting stitches, the favorites of which have been collected into stitch treasuries. A piece of knitting begins with the process of *casting on* (also known as "binding on"), which involves the initial creation of the stitches on the needle. Different methods of casting on are used for different effects: one may be stretchy enough for lace, while another provides a decorative edging — *Provisional* cast-ons are used when the knitting will continue in both directions from the cast-on. There are various method employed to "cast on," such as the "thumb method" (also known as "slingshot" or "long-tail" cast-ons), where the stitches are created by a series of loops that will, when knitted, give a very loose edge ideal for "picking up stitches" and knitting a border; the "double needle method" (also known as "knit-on" or "cable cast-on"), whereby each loop placed on the needle is then "knitted on," which produces a firmer edge ideal on its own as a border; and many more. The number of active stitches remains the same as when cast on unless stitches are added (an increase) or removed (a decrease).

Most Western-style knitters follow either the English style (in which the yarn is held in the right hand) or the Continental style (in which the yarn is held in the left hand). A third but less common method, called combination knitting, may also be used.

Once the knitted piece is finished, the remaining live stitches are "cast off." Casting (or "binding") off loops the stitches across each other so they can be removed from the needle without unravelling the item. Although the mechanics are different from casting on, there is a similar variety of methods.

In knitting certain articles of clothing, especially larger ones like sweaters, the final knitted garment will be made of several knitted pieces, with individual sections of the garment knit separately and then sewn together. Seamless knitting, where a whole garment is knit as a single piece, is also possible. Elizabeth Zimmermann is probably the best-known proponent of seamless or circular knitting techniques. Smaller items, such as socks and hats, are usually knit in one piece on double-pointed needles or circular needles.

Social Aspects

This woman is knitting at a coffee shop; although it can be done by one person alone, knitting can be a social activity. There are many knitting guilds and other knitting groups or knitting clubs.

One of the earliest known examples of knitting was finely decorated cotton socks found in Egypt in the end of the first millennium AD. The first knitting trade guild was started in Paris in 1527. With the invention of the knitting machine, however, knitting "by hand" became a useful but non-essential craft. Similar to quilting, spinning, and needlepoint, knitting became a social activity.

Hand knitting has gone into and out of fashion many times in the last two centuries, and at the turn of the 21st century it is enjoying a revival. According to the industry group Craft Yarn Council of America, the number of women knitters in the United States age 25–35 increased 150% in the two years between 2002 and 2004. While some may say hand knitting has never really gone away, this latest reincarnation is less about the make do and mend of the 1940s and 50's and more about making a statement about individuality as well as developing an innate sense of community.

Additionally, many contemporary knitters have an interest in blogging about their knitting, knitting patterns, and techniques, or joining a virtual community focused on knitting, such as the extremely popular Ravelry. There are also a number of popular knitting podcasts, and various other knitting websites. One knitting website, loveknitting, allows users to upload pictures of projects and add tags. Other users can comment, encouraging conversation and continuous learning. Contemporary knitting groups may be referred to in the U.S. as a "Stitch 'N Bitch" where a group of knitters get together to work on projects, discuss [knitting patterns], troubleshoot their work and just socialize. In the UK, the term has been "knitting circle" since the early 20th century.

There are now numerous groups that are not only growing individually, but also forming international communities. Communities also exist online, with blogs being very popular, alongside online groups and social networking through mediums such as Yahoo! Groups, where people can share tips and techniques, run competitions, and share their knitting patterns. More people are finding knitting a recreation and enjoying the hobby with their family. Knitting parties also are becoming popular in small and large communities around the U.S. and Canada.

Graffiti

In the last decade, a practice called knitting graffiti, guerilla knitting, or yarn bombing— the use of knitted or crocheted cloth to modify and beautify one's (usually outdoor) surroundings—emerged in the U.S. and spread worldwide. Magda Sayeg is credited with starting the movement in the US and Knit the City are a prominent group of graffiti knitters in the United Kingdom. Yarn bombers sometimes target existing pieces of graffiti for beautification. For instance, Dave Cole is a contemporary sculpture artist who practiced knitting as graffiti for a large-scale public art installation in Melbourne, Australia for the Big West Arts Festival in 2009. The work was vandalized the night of its completion. A new movie, shot by a Tasmanian filmmaker on a set made almost entirely out of yarn, was partially inspired by "knitted graffiti".

Charity

Knitting garments for free distribution to others is a common theme in modern history.

Knitters made socks, sweaters, scarves, mittens, gloves, and hats for soldiers in Crimea, the American Civil War, and the Boer Wars; this practice continued in World War I, World War II and the Korean War, and continues for soldiers in Iraq and Afghanistan. In these historical projects, yarn companies often provided knitting patterns approved by the various branches of the armed services; they were distributed by local chapters of the American Red Cross or other organizations. Modern projects usually entail the knitting of hats or helmet liners; the liners provided for soldiers must be of 100% worsted weight wool and be crafted using specific colors.

The Australian charity *Wrap with Love* provides blankets knitted by volunteers to people most in need around the world who have been affected by war. Clothing and afghans are frequently made for children, the elderly, and the economically disadvantaged in various countries. Pine Ridge Indian Reservation accepts donations for the Lakota people in the United States. Prayer shawls, or shawls in which the crafter meditates or says prayers of their faith while knitting with the intent on comforting the recipient, are donated to those experiencing loss or stress. Many knitters today knit and donate "chemo caps," soft caps for cancer patients who lose their hair during chemotherapy. Yarn companies offer free knitting patterns for these caps.

Originally started after the 2004 Indonesian tsunami, Knitters Without Borders is a charity challenge issued by knitting personality Stephanie Pearl-McPhee that encourages knitters to donate to Médecins Sans Frontières (Doctors Without Borders). Instead of knitting for charity, knitters are encouraged to donate a week's worth of disposable income, including money that otherwise might have been spent on yarn. Knitted items are occasional offered as prizes to donors. As of September 2011, Knitters Without Borders donors have contributed CAD$1,062,217.

Security blankets can also be made through the Project Linus organization which helps needy children.

There are organizations that help reach other countries in need such as afghans for Afghans. This outreach is described as, "afghans for Afghans is a humanitarian and educational people-to-people project that sends hand-knit and crocheted blankets and sweaters, vests, hats, mittens, and socks to the beleaguered people of Afghanistan."

Penguin sweaters were knitted by volunteers for the rehabilitation of penguins contaminated by exposure to oil slicks. The project is now complete.

Chicken sweaters were also knitted to aid battery hens that had lost their feathers. The organization is not currently accepting donations, but maintains a list of volunteers.

Sweater *Curse*

Knitting especially large or fine garments such as sweaters can require months of work and, as gifts, may have a strong emotional aspect. The so-called *sweater curse* expresses the experience that a significant other will break up with the knitter immediately after receiving a costly hand-knit gift such as a sweater. A significant minority of knitters claim to have experienced the sweater curse; a recent poll indicated that 15% of active knitters say they have experienced the sweater curse firsthand, and 41% consider it a possibility that should be taken seriously. Although sometimes labeled a "superstition", the sweater curse is not treated in knitting literature as anything paranormal.

Psychological and Meditative Aspects

The oral histories of many knitters have been collected, and suggest that hand-knitting is often associated with compassion. "I knit love into every stitch" is a common refrain.

The repetitive aspect of hand-knitting is generally relaxing and can be well-suited for meditational or spiritual practice.

Health Benefits

Studies have shown that knitting, along with other forms of needlework, provide several significant health benefits. These studies have found the rhythmic and repetitive action of knitting can "help prevent and manage stress, pain and depression, which in turn strengthens the body's immune system", as well as create a relaxation response in the body which can decrease blood pressure, heart rate, help prevent illness, and have a calming effect. Pain specialists have also found that the brain chemistry is changed when one knits, resulting in an increase in "feel good" hormones (i.e. serotonin and dopamine), and a decrease in stress hormones.

Knitting, along with other leisure activities has been linked to reducing the risk of developing Alzheimer's disease and dementia. Much like physical activity strengthens the body, mental exercise makes the human brain more resilient.

A repository of research into the effect on health of knitting can be found at Stitchlinks, an organisation founded in Bath, England.

Mega Knitting

Mega knitting is hand knitting using knitting needles greater than or equal to half an inch in diameter.

Mega knitting uses the same stitches and techniques as conventional knitting, except that hooks may be carved into the ends of the needles. The hooked needles greatly enhance control of the work, catching the stitches and preventing them from slipping off.

Mega knitting produces a chunky, bulky fabric or an open lacy weave, depending on the weight and type of yarn used.

The current holder of the Guinness World Record for Knitting with the Largest Knitting Needles is Julia Hopson of Penzance in Cornwall.Julia knitted a square of ten stitches and ten rows in stockinette stitch using knitting needles that were 6.5 centimeters in diameter and 3.5 meters long.

In Literature

Knitting is sometimes featured in literature. Knitting and its techniques may be used as a metaphor; its meditative and spiritual aspects may be emphasized; it may signal various types of domesticity; or it may be used for dramatic irony, as when an apparently harmless knitter proves deadly and implacable. Examples from 19th century novels include Madame Thérèse Defarge in Charles Dickens' *A Tale of Two Cities*, Anna Makarovna in Leo Tolstoy's *War and Peace*, various characters in Jane Austen's novels and Miss Ophelia in Harriet Beecher Stowe's *Uncle Tom's Cabin*. Several characters in Virginia Wolff's novels are knitters. In the first decade of the 21st century, knitting has been a key element in several novels and even murder mysteries.

Knitting Materials

Yarn

Yarn for hand-knitting is usually sold as balls or skeins (hanks), although it may also be wound on spools or cones. Skeins and balls are generally sold with a *yarnband*, a label that describes the yarn's weight, length, dye lot, fiber content, washing instructions, suggested needle size, likely gauge, etc. It is common practice to save the yarn band for future reference, especially if additional skeins must be purchased. Knitters generally ensure that the yarn for a project comes from a single dye lot. The dye lot specifies a group of skeins that were dyed together and thus have precisely the same color; skeins from different dye-lots, even if very similar in color, are usually slightly different and may produce a visible stripe when knitted into the same project. If a knitter buys insufficient yarn of a single dye lot to complete a project, additional skeins of the same dye lot can sometimes be obtained from other yarn stores or online.

The thickness of the yarn is a significant factor in determining the gauge, i.e., how many stitches and rows are required to cover a given area for a given stitch pattern. Thicker yarns generally require thicker knitting needles, whereas thinner yarns may be knit with thick or thin needles. Hence, thicker yarns generally require fewer stitches, and therefore less time, to knit up a given garment. Patterns and motifs are coarser with thicker yarns; thicker yarns produce bold visual effects, whereas thinner yarns are best for refined patterns. Yarns are grouped by thickness into six categories: superfine, fine, light, medium, bulky and superbulky; quantitatively, thickness is measured by the number of wraps per inch (WPI). The related *weight per unit length* is usually measured in tex or dernier.

In addition to choosing the correct thickness for the gauge, the knitter must also pick the type of yarn fiber. There are currently about fifteen types of yarn fiber, falling into two categories, natural and synthetic. Natural fibers are those that are obtained from a plant or an animal and have different attributes depending on the animal/plant they are harvested from which must be taken into account when considering the uses of a finished knitting object. Example: Wool is well suited to items which will be used to hold in heat, even when damp, such as winter hats and mittens. Linen, however, would be well suited to a light summer sweater when breath ability is a factor. Synthetic fibers are made by forcing a thick solution of polymerized chemicals through spinneret nozzles and hardening the resulting filament in a chemical bath. Natural fibers are generally softer and more comfortable whereas synthetics are durable and easier to dye. Some fibers can be harder to knit with than others for a variety of reasons. cotton, for example, doesn't stretch as much as wool, and as such requires the knitter to work harder to maintain gauge.

Before knitting, the knitter will typically transform a hank into a ball where the yarn emerges from the center of the ball; this making the knitting easier by preventing the yarn from becoming easily tangled. This transformation may be done by hand, or with a device known as a ballwinder. When knitting, some knitters enclose their balls in jars to keep them clean and untangled with other yarns; the free yarn passes through a small hole in the jar-lid.

A yarn's usefulness for a knitting project is judged by several factors, such as its *loft* (its ability to trap air), its *resilience* (elasticity under tension), its washability and colorfastness, its *hand* (its feel, particularly softness vs. scratchiness), its durability against abrasion, its resistance to pilling, its *hairiness* (fuzziness), its tendency to twist or untwist, its overall weight and drape, its blocking and felting qualities, its comfort (breathability, moisture absorption, wicking properties) and of course its look, which includes its color, sheen, smoothness and ornamental features. Other factors include allergenicity; speed of drying; resistance to chemicals, moths, and mildew; melting point and flammability; retention of static electricity; and the propensity to become stained and to accept dyes. Different factors may be more significant than others for different knitting projects, so there is no one "best" yarn. The resilience and propensity to (un)twist are general properties that affect the ease of hand-knitting. More resilient yarns are more forgiving of irregularities in tension; highly twisted yarns are sometimes difficult to knit, whereas untwisting yarns can lead to split stitches, in which not all the yarn is knitted into a stitch. A key factor in knitting is *stitch definition*, corresponding to how well complicated stitch patterns can be seen when made from a given yarn. Smooth, highly spun yarns are best for showing off stitch patterns; at the other extreme, very fuzzy yarns or eyelash yarns have poor stitch definition, and any complicated stitch pattern would be invisible.

Tools

The process of knitting has three basic tasks: (1) the active (unsecured) stitches must be held so they don't drop; (2) these stitches must be released sometime after they are secured; and (3) new bights of yarn must be passed through the fabric, usually through active stitches, thus securing them. In very simple cases, knitting can be done without tools, using only the fingers to do these tasks; however, hand-knitting is usually carried out using tools such as knitting needles or rigid frames. Depending on their size and shape, the rigid frames are called knitting boards, knitting rings (also called knitting looms) or knitting spools (also known as knitting knobbies, knitting nancies, or corkers). Other tools are used to prepare yarn for knitting, to measure and design knitted garments, or to make knitting easier or more comfortable.

Needles

There are four basic types of knitting needles (also called "knitting pins"). The first and most common type consists of two slender, straight sticks tapered to a point at one end, and with a knob at the other end to prevent stitches from slipping off. Such needles are usually 10-16 inches long but, due to the compressibility of knitted fabrics, may be used to knit pieces significantly wider. The most important property of needles is their diameter, which ranges from below 2 mm to 25 mm (roughly 1 inch). The diameter affects the size of stitches, which affects the gauge of the knitting and the elasticity of the fabric. Thus, a simple way to change gauge is to use different needles, which is the basis of uneven knitting. Although knitting needle diameter is often measured in millimeters, there are several different size systems, particularly those specific to the United States, the United Kingdom and Japan; a conversion table is given in the knitting needle article. Such knitting needles may be made out of any materials, but the most common materials are metals, wood, bamboo, and plastic. Different materials have different frictions and grip the yarn differently; slick needles such as metallic needles are useful for swift knitting, whereas rougher needles such as bamboo are less prone to dropping stitches. The knitting of new stitches occurs only at the tapered ends, and needles with lighted tips have been sold to allow knitters to knit in the dark.

The second type of knitting needles are straight, double-pointed knitting needles (also called "dpns"). Double-pointed needles are tapered at both ends, which allows them to be knit from either end. Dpns are typically used for circular knitting, especially smaller tube-shaped pieces such as sleeves, collars, and socks; usually one needle is active while the others hold the remaining active stitches. Dpns are somewhat shorter (typically 7 inches) and are usually sold in sets of four or five.

Cable needles are a special case of dpns, although they usually are not straight, but dimpled in the middle. Cable needles are typically very short (a few inches), and are used to hold stitches temporarily while others are being knitted. Cable patterns are made by permuting the order of stitches; although one or two stitches may be held by hand or knit out of order, cables of three or more generally require a cable needle.

The third needle type consists of circular needles, which are long, flexible double-pointed needles. The two tapered ends (typically 5 inches (130 mm) long) are rigid and straight, allowing for easy knitting; however, the two ends are connected by a flexible strand (usually nylon) that allows the two ends to be brought together. Circular needles are typically 24-60 inches long, and are usually used singly or in pairs; again, the width of the knitted piece may be significantly longer than the length of the circular needle. Special kits are available that allow circular needles of various lengths and diameters to be made as needed; rigid ends of various diameters may be screwed into strands of various lengths. The ability to work from either end of one needle is convenient in several types of knitting, such as slip-stitch versions of double knitting. Circular needles may be used for flat or circular knitting.

The fourth type of needle is a hybrid needle. It is a straight needle with a point on one end and a flexible strand on the other end with a stopper, such as a large lightweight bead, at the end. This type of needle allows a larger project to be worked at one time than a straight needle, while folding up quickly and more compactly for travel.

Mega Needles

The largest aluminum circular knitting needles on record are size US 150 and are nearly 7 feet tall. They are owned by Paradise Fibers and are currently on display in the Paradise Fibers retail showroom.

Ancillary Tools

Various tools have been developed to make hand-knitting easier. Tools for measuring needle diameter and yarn properties have been discussed above, as well as the yarn swift, ballwinder and "yarntainers". Crochet hooks and a darning needle are often useful in binding off or in joining two knitted pieces edge-to-edge. The darning needle is used in duplicate stitch (also known as Swiss darning), while the crochet hook is also essential for repairing dropped stitches and some specialty stitches such as tufting. Other tools such as knitting spools or pom-pom makers are used to prepare specific ornaments. For large or complex knitting patterns, it is sometimes difficult to keep track of which stitch should be knit in a particular way; therefore, several tools have been developed to identify the number of a particular row or stitch, including circular stitch markers, hanging markers, extra yarn and counters. A second potential difficulty is that the knitted piece will slide off the tapered end of the needles when unattended; this is prevented by "point protectors" that cap the tapered ends. Another problem is that too much knitting may lead to hand and wrist troubles; for this, special stress-relieving gloves are available. Finally, there are sundry bags and containers for holding knitting projects, yarns and needles.

Knitted Fabric

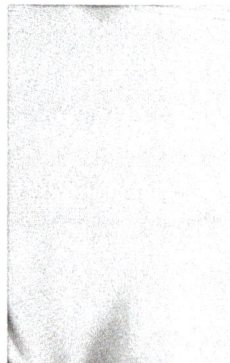

A knitted fabric cloth sample

Knitted fabric is a textile that results from knitting. Its properties are distinct from woven fabric in that it is more flexible and can be more readily constructed into smaller pieces, making it ideal for socks and hats.

Its properties are distinct from nonwoven fabric in that it is more durable but takes more resources to create, making it suitable for multiple uses.

Weft- and Warp-knit Fabric

There are two major varieties of knit fabric: weft-knit and warp-knit fabrics. Warp-knitted fabrics such as tricot and milanese are resistant to runs, and are commonly used in lingerie.

Weft-knit fabrics are easier to make and more common. When cut, they will unravel (run) unless repaired.

Warp-knit fabrics are resistant to runs and relatively easy to sew. Raschel lace—the most common type of machine made lace—is a warp knit fabric but using many more guide-bars (12+) than the usual machines which mostly have three or four bars.

Structure of Knitted Fabrics

Courses and Wales

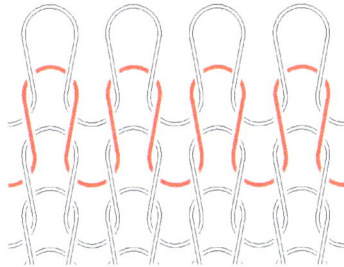

Structure of stockinette, a common weft-knit fabric. The meandering red path defines one *course*, the path of the yarn through the fabric. The uppermost white loops are unsecured and "active", but they secure the red loops suspended from them. In turn, the red loops secure the white loops just below them, which in turn secure the loops below them, and so on.

Alternating wales of red and white knit stitches. Each stitch in a wale is suspended from the one above it.

In weaving, threads are always straight, running parallel either lengthwise (warp threads) or crosswise (weft threads). By contrast, the yarn in knitted fabrics follows a meandering path (a *course*), forming symmetric loops (also called bights) symmetrically above and below the mean path of the yarn. These meandering loops can be easily stretched in different directions giving knit fabrics much more elasticity than woven fabrics. Depending on the yarn and knitting pattern, knitted garments can stretch as much as 500%. For this reason, knitting is believed to have been developed for garments that must be elastic or stretch in response to the wearer's motions, such as socks and hosiery. For comparison, woven garments stretch mainly along one or other of a related pair of directions that lie roughly diagonally between the warp and the weft, while contracting in the other direction of the pair (stretching and contracting with the *bias*), and are not very elastic, unless they are woven from stretchable material such as spandex. Knitted garments are often more form-fitting than woven garments, since their elasticity allows them to contour to the body's outline more closely; by contrast, curvature is introduced into most woven garments only with sewn darts, flares, gussets and gores, the seams of which lower the elasticity of the woven fabric still further. Extra curvature can be introduced into knitted garments without seams, as in the heel of a sock; the effect of darts, flares, etc. can be obtained with short rows or by increasing or decreasing the number of stitches. Thread used in weaving is usually much finer than the yarn used in knitting, which can give the knitted fabric more bulk and less drape than a woven fabric.

If they are not secured, the loops of a knitted course will come undone when their yarn is pulled; this is known as *ripping out*, *unravelling* knitting, or humorously, *frogging* (because you 'rip it', this sounds like a frog croaking: 'rib-bit'). To secure a stitch, at least one new loop is passed through it. Although the new stitch is itself unsecured ("active" or "live"), it secures the stitch(es) suspended from it. A sequence of stitches in which each stitch is suspended from the next is called a *wale*. To secure the initial stitches of a knitted fabric, a method for casting on is used; to secure the final stitches in a wale, one uses a method of binding/casting off. During knitting, the active stitches are secured mechanically, either from individual hooks (in knitting machines) or from a knitting needle or frame in hand-knitting.

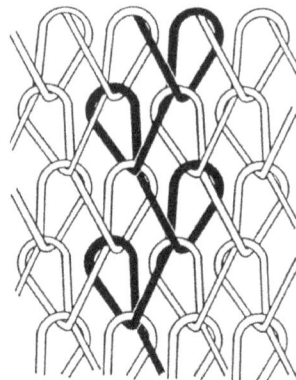

Basic pattern of warp knitting. Parallel yarns zigzag lengthwise along the fabric,
each loop securing a loop of an adjacent strand from the previous row.

Knitting Stitches and Stitch Patterns

Different stitches and stitch combinations affect the properties of knitted fabric. Individual stitches look differently; knit stitches look like "V"'s stacked vertically, whereas purl stitches look like a wavy horizontal line across the fabric. Patterns and pictures can be created using colors in knitted fabrics by using stitches as "pixels"; however, such pixels are usually rectangular, rather than square. Individual stitches, or rows of stitches, may be made taller by drawing more yarn into the new loop (an elongated stitch), which is the basis for uneven knitting: a row of tall stitches may alternate with one or more rows of short stitches for an interesting visual effect. Short and tall stitches may also alternate within a row, forming a fish-like oval pattern.

Stitches also affect the physical properties of a fabric. Stockinette stitch forms a smooth nap. Aran knitting patterns are used to create a bulkier fabric to retain heat.

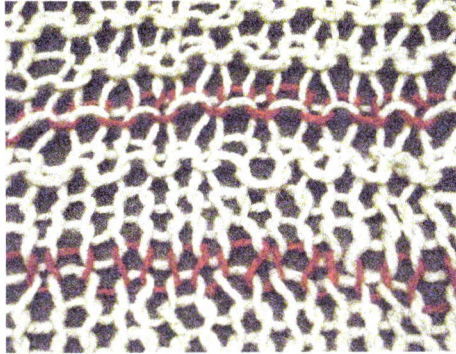

Two courses of red yarn illustrating two basic fabric types. The lower red course is knit into the white row below it and is itself knit on the next row; this produces *stockinette* stitch. The upper red course is purled into the row below and then is knit, consistent with *garter* stitch.

A dropped stitch, or missed stitch, is a common error that creates an extra loop to be fixed.

In the simplest knitted fabric pattern, all the stitches are knit or purl; this is known as a garter stitch. Alternating rows of knit stitches and purl stitches produce what is known as a stockinette pattern/stocking stitch. Vertical stripes (ribbing) are possible by having alternating wales of knit and purl stitches. For example, a common choice is 2x2 ribbing, in which two wales of knit stitches are followed by two wales of purl stitches, etc. Horizontal striping (welting) is also possible, by alternating *rows* of knit and purl stitches. Checkerboard patterns (basketweave) are also possible, the smallest of which is known as *seed/moss stitch*: the stitches alternate between knit and purl in every wale and along every row.

Fabrics in which the number of knit and purl stitches are not the same, such as stockinette/stocking stitch, have a tendency to curl; by contrast, those in which knit and purl stitches are arranged symmetrically (such as ribbing, garter stitch or seed/moss stitch) tend to lie flat and drape well. Wales of purl stitches have a tendency to recede, whereas those of knit stitches tend to come forward. Thus, the purl wales in ribbing tend to be invisible, since the neighboring knit wales come forward. Conversely, rows of purl stitches tend to form an embossed ridge relative to a row of knit stitches. This is the basis of shadow knitting, in which the appearance of a knitted fabric changes when viewed from different directions.

The stitches on the right are right-plaited, whereas the stitches on the left are left-plaited.

Right- and Left-plaited Stitches

Both types of plaited stitches give a subtle but interesting visual texture, and tend to draw the fabric inwards, making it stiffer. Plaited stitches are a common method for knitting jewelry from fine metal wire.

Illustration of entrelac. The blue and white wales are parallel to each other, but both are perpendicular to the black and gold wales, resembling basket weaving.

Edges and Joins Between Fabrics

The initial and final edges of a knitted fabric are known as the *cast-on* and *bound/cast-off* edges. The side edges are known as the *selvages*; the word derives from "self-edges", meaning that the stitches do not need to be secured by anything else. Many types of selvages have been developed, with different elastic and ornamental properties.

Edges are introduced within a knitted fabric for button holes, pockets, or decoration, by binding/casting off and re-casting on again (horizontal) or by knitting the fabrics on either side of an edge separately.

Two knitted fabrics can be joined by embroidery-based grafting methods, most commonly the Kitchener stitch. New wales can be begun from any of the edges of a knitted fabric; this is known as picking up stitches and is the basis for entrelac, in which the wales run perpendicular to one another in a checkerboard pattern.

Illustration of cable knitting. The central braid is formed from 2x2 ribbing in which the background is formed of purl stitches and the cables are each two wales of knit stitches. By changing the order in which the stitches are knit, the wales can be made to cross.

Cables, Increases, and Lace

When knit wales cross, a cable is formed. Cables patterns tend to draw the fabric together, making it denser and less elastic; Aran sweaters are a common form of knitted cabling. Arbitrarily complex braid patterns can be done in cable knitting.

Lace knitting consists of making patterns and pictures using holes in the knit fabric, rather than with the stitches themselves. The large and many holes in lacy knitting makes it extremely elastic; for example, some Shetland "wedding-ring" shawls are so fine that they may be drawn through a wedding ring.

In lace knitting, a pattern is formed by making small, stable holes in the fabric.

By combining increases and decreases, it is possible to make the direction of a wale slant away from vertical, even in weft knitting. This is the basis for bias knitting, and can be used for visual effect, similar to the direction of a brush-stroke in oil painting.

Ornamentations and Additions

Various point-like ornaments may be added to a knit fabric for their look or to improve the wear of the fabric. Examples include various types of bobbles, sequins and beads. Long loops can also be drawn out and secured, forming a "shaggy" texture to the fabric; this is known as loop knitting. Additional patterns can be made on the surface of the knitted fabric using embroidery; if the embroidery resembles knitting, it is often called Swiss darning. Various closures for the garments, such as frogs and buttons can be added; usually buttonholes are knitted into the garment, rather than cut.

Ornamental pieces may also be knitted separately and then attached using applique. For example, differently colored leaves and petals of a flower could be knit separately and attached to form the final picture. Separately knitted tubes can be applied to a knitted fabric to form complex Celtic knots and other patterns that would be difficult to knit.

Unknitted yarns may be worked into knitted fabrics for warmth, as is done in tufting and "weaving" (also known as "couching").

Properties of Fabrics

The topology of a knitted fabric is relatively complex. Unlike woven fabrics, where strands usually run straight horizontally and vertically, yarn that has been knitted follows a looped path along its row, as with the red strand in the diagram at left, in which the loops of one row have all been pulled through the loops of the row below it.

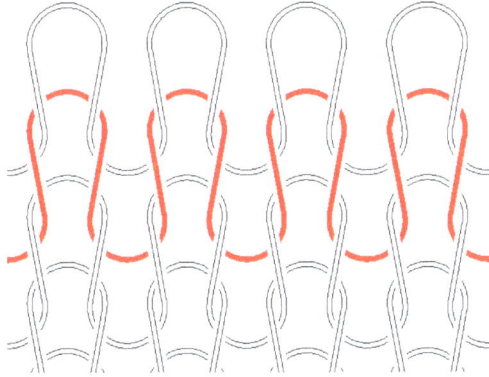

Schematic of stockinette stitch, the most basic weft-knit fabric

Because there is no single straight line of yarn anywhere in the pattern, a knitted piece of fabric can stretch in all directions. This elasticity is all but unavailable in woven fabrics which only stretch along the bias. Many modern stretchy garments, even as they rely on elastic synthetic materials for some stretch, also achieve at least some of their stretch through knitted patterns.

Close-up of front of stockinette stitch

Close-up of back of stockinette stitch, also same appearance as reverse stockinette stitch

The basic knitted fabric (as in the diagram, and usually called a *stocking* or *stockinette* pattern) has a definite "right side" and "wrong side". On the right side, the visible portions of the loops are the verticals connecting two rows which are arranged in a grid of *V* shapes. On the wrong side, the ends of the loops are visible, both the tops and bottoms, creating a much more bumpy texture sometimes called *reverse stockinette*. (Despite being the "wrong side," reverse stockinette is frequently used as a pattern in its own right.) Because the yarn holding rows together is all on the front, and the yarn holding side-by-side stitches together is all on the back, stockinette fabric has a strong tendency to curl toward the front on the top and bottom, and toward the back on the left and right side.

Stitches can be worked from either side, and various patterns are created by mixing regular knit stitches with the "wrong side" stitches, known as purl stitches, either in columns (ribbing), rows (garter, welting), or more complex patterns. Each fabric has different properties: a garter stitch has much more vertical stretch, while ribbing stretches much more horizontally. Because of their front-back symmetry, these two fabrics have little curl, making them popular as edging, even when their stretch properties are not desired.

Different combinations of knit and purl stitches, along with more advanced techniques, generate fabrics of considerably variable consistency, from gauzy to very dense, from highly stretchy to relatively stiff, from flat to tightly curled, and so on.

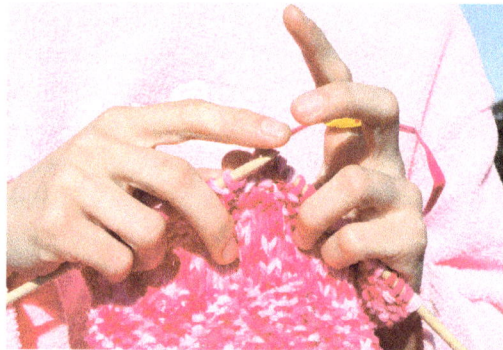

Close-up of knitting

Texture

The most common texture for a knitted garment is that generated by the flat stockinette stitch—as seen, though very small, in machine-made stockings and T-shirts—which is worked in the round as nothing but knit stitches, and worked flat as alternating rows of knit and purl. Other simple textures can be made with nothing but knit and purl stitches, including garter stitch, ribbing, and moss and seed stitches. Adding a "slip stitch" (where a loop is passed from one needle to the other) allows for a wide range of textures, including heel and linen stitches as well as a number of more complicated patterns.

Close-up of ribbing

Some more advanced knitting techniques create a surprising variety of complex textures. Combining certain increases, which can create small eyelet holes in the resulting fabric, with assorted decreases is key to creating knitted lace, a very open fabric resembling lace. Open vertical stripes can be created using the drop-stitch knitting technique. Changing the order of stitches from one row to the next, usually with the help of a cable needle or stitch holder, is key to cable knitting, producing an endless variety of cables, honeycombs, ropes, and Aran sweater patterning. Entrelac forms a rich checkerboard texture by knitting small squares, picking up their side edges, and knitting more squares to continue the piece.

Fair Isle knitting uses two or more colored yarns to create patterns and forms a thicker and less flexible fabric.

The appearance of a garment is also affected by the *weight* of the yarn, which describes the thickness of the spun fibre. The thicker the yarn, the more visible and apparent stitches will be; the thinner the yarn, the finer the texture.

Color

Plenty of finished knitting projects never use more than a single color of yarn, but there are many ways to work in multiple colors. Some yarns are dyed to be either *variegated* (changing color every few stitches in a random fashion) or *self-striping* (changing every few rows). More complicated techniques permit large fields of color (intarsia, for example), busy small-scale patterns of color (such as Fair Isle), or both (double knitting and slip-stitch color, for example).

Yarn with multiple shades of the same hue are called *ombre*, while a yarn with multiple hues may be known as a given *colorway* — a green, red and yellow yarn might be dubbed the "Parrot Colorway" by its manufacturer, for example. *Heathered* yarns contain small amounts of fibre of different colours, while *tweed* yarns may have greater amounts of different colored fibres.

Composition of Knitted Fabrics

The most common fibres used for knitted fabrics are cotton & viscose with or without elastane, these tend to be single jersey construction and are used for most t-shirt style tops.

Knitted dress, 1967

History of Fashion Knitwear

Coco Chanel's 1916 use of jersey in her hugely influential suits was a turning point for knitwear, which became associated with the woman. Shortly afterwards, Jean Patou's cubist-inspired, color-blocked knits were the sportswear of choice.

In the 1940s came the iconic wearing of body-skimming sweaters by sex symbols like Lana Turner and Jane Russell, though the 1950s were dominated by conservative pop-corn knits. The swinging 1960s were famously manifested in Missoni's colorful zigzag knitwear. This era also saw the rise both of Sonia Rykiel, dubbed the "Queen of Knit-wear" for her vibrant striped sweaters and her clingy dresses, and of Kennedy-inspired preppy sweaters.

In the 1980s, knitwear emerged from the realm of sportswear to dominate high fashion; notable designs included Romeo Gigli's "haute-bohemian cocoon coats" and Ralph Lauren's floor-length cashmere turtlenecks.

Contemporary knitwear designers include Diane von Furstenberg, and Irakli Nasidze.

Basic Knitted Fabrics

Basic knitted fabrics are so fundamental that some types have been adopted as part of the language of knitting, similar to techniques such as yarn over or decrease. Examples include stockinette stitch, reverse stockinette stitch, garter stitch, seed stitch, faggoting, and tricot. In some cases, these fabrics appear differently on the right side (as seen when making the stitch) than on the wrong side (as seen from the other side, when the work is turned).

Stockinette/Stocking Stitch and Reverse Stockinette Stitch

Stockinette stitch (in UK, Australia, New Zealand etc., stocking stitch) is the most basic knitted fabric; every stitch is a knit stitch. In the round, stockinette stitch is produced by knitting every stitch; by contrast, in the flat, stockinette stitch is produced by knitting and purling alternate rows.

Stockinette-stitch fabric is very smooth and each column ("wale") resembles a stacked set of "V"'s. It has a strong tendency to curl horizontally and vertically because of the asymmetry of its faces.

Reverse stockinette stitch is produced in the same way as stockinette, except that the purl stitches are done on the right side and the knit stitches on the wrong side. In the round, reverse stockinette stitch is produced by purling every stitch.

Stockinette stitch front.

Back of stockinette stitch, also same appearance as reverse stockinette stitch.

Garter Stitch

Garter stitch

Garter stitch, also known as the knit stitch, is the most basic form of welting (as seen from the right side). In the round, garter stitch is produced by knitting and purling alternate rows. By contrast, in the flat, garter stitch is produced by knitting every stitch (or purling every stitch, though this is much less common).

In garter-stitch fabrics, the "purl" rows stand out from the "knit" rows, which provides the basis for shadow knitting. Garter-stitch fabric has significant lengthwise elasticity and little tendency to curl, due to the symmetry of its faces.

Seed/Moss Stitch

Seed stitch (called moss stitch in the UK, Australia, New Zealand etc.) is the most basic form of a basketweave pattern; knit and purl stitches alternate in every column ("wale") *and* every row ("course"). In other words, every knit stitch is flanked on all four sides (left and right, top and bottom) by purl stitches, and vice versa.

Seed/moss-stitch fabrics lie flat; the symmetry of their two faces prevents them from curling to one side or the other. Hence, it makes an excellent choice for edging, e.g., the central edges of a cardigan. However, seed stitch is "nubbly", not nearly as smooth as stockinette/stocking stitch.

Faggoting

Faggoting lace

Faggoting is a variation of lace knitting, in which every stitch is a yarn over or a decrease. There are several types of faggoting, but all are an extremely open lace similar to netting.

Like most lace fabrics, faggoting has little structural strength and deforms easily, so it has little tendency to curl despite being asymmetrical. Faggoting is stretchy and open, and most faggoting stitches look the same on both sides, making them ideal for garments like lacy scarves or stockings.

Tricot Knitting

Tricot is a special case of warp knitting, in which the yarn zigzags vertically, following a single *column* ("wale") of knitting, rather than a single *row* ("course"), as is customary. Tricot and its relatives are very resistant to runs, and are commonly used in lingerie.

Other Basic Fabrics

Other classes of basic knitted fabrics include ribbing, welting and cables.

Knitting Technology

Knitting is the process of fabric formation by producing series of intermeshed loops. Loops are the building blocks of knitted fabrics. As a result, the knitted fabrics, in general, are more stretchable than the woven fabrics. The open structure of knitted fabrics also helps in moisture vapour transmission making it suitable for the sports garments. Besides, the knitted fabrics have more porosity than the woven fabrics. Therefore, knitted structures can trap more air resulting in lower thermal conductivity and higher thermal resistance. There are two types of principles of knitting namely warp knitting and weft knitting.

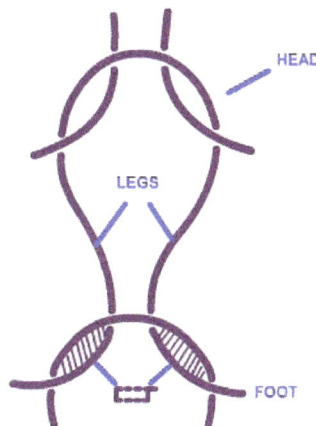

A knitted loop

Weft Knitting

In this method loops are made by each weft yarn and loops are formed across the width of the fabric (figure a). Weft knitted structure can be made even from one supply package. The weft knitting machines are of two types.

☐ Flat bed machine

- Single bed

- Double bed or V bed

☐ Circular bed knitting machine

In flat bed machine, the needles do not perform any lateral movement. The axial movement of the needles, needed for loop formation, is actuated by a set of cams mounted on cam jacket which reciprocate laterally (exception: straight bar machines). In contrast, the cam jackets are generally stationary in circular knitting machine. The cylinder carrying the needles on its grooved surface rotates continuously to cause the upward and downward movement of needles. In many small diameter circular weft knitting machines, the cylinder may remain stationary while the cam jackets revolve. This is true for single feeder machines.

Warp Knitting

In this method loops are made from each warp yarn and loops are formed along the length of the fabric (figure b). The yarns are supplied in the form of sheet made by parallel warp yarns coming out from a single or multiple warp beams. The yarns are fed to the needles by guide bars which swings to and fro and shog laterally. The loop formation mechanism is more complex for warp knitting.

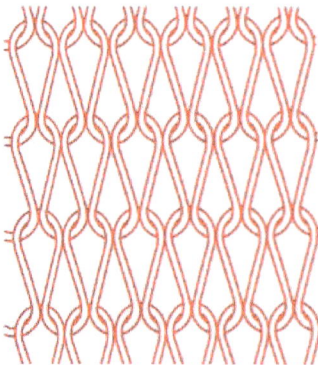

(a) Weft knitted fabric   (b) Warp knitted fabric

Weft and warp knitted fabrics

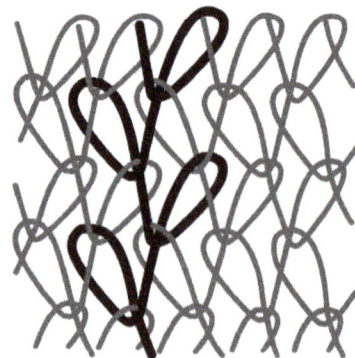

Needle

Irrespective of the knitting technology, the machine element which helps in forming the loop is called needle. Latch, bearded and compound needles are used depending on the type of knitting machine. Latch needle is most popularly used in weft knitting and Raschel warp knitting machine.

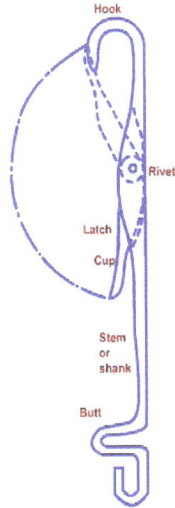

Latch needle

The major components of a latch needle are

- Hook
- Latch
- Latch spoon (cup)
- Stem
- Butt

Hook is the curved part of the needle which is responsible for forming the loop. Latch is a tiny component and it is riveted on the upper part of the stem of the needle. Latch spoon is the tip of the latch which encloses the tip of the hook when the latch closes. The upward and downward movements of the needles during the loop formation are caused by a set of cams in weft knitting and by movement of bars in warp knitting . The butt is actually the 'follower' and it is pressed against the cam to cause movement of the needle.

Loop Formation in Knitting

The sequence of loop formation is shown in the figure. When the needle moves up, the old loop forces the latch to open. When the old loop rests on the latch, the position is called 'tuck' position (1 in figure). Then the needle moves up further and the old loop slides down the latch and rest on the stem of the needle. This is called the 'clearing' position (2 in figure). The needle attains its highest position at 3 in figure. Then the needle starts to descend and the hook catches the yarn. As the needle continues to descend, the yarn bends in the form of a loop (U shape). The old loop now helps to close the latch by pushing it in upward direction so that newly formed loop is caught between the hook and latch (4 in figure). The needle continues to descend and new loop is 'cast on' (5 in figure) and finally 'knocked over' (6 in figure) through the old one. Casting off or knocking over is the same phenomena, executed in two different manners. For casting-off to take place special knitting elements bodily push the old loop out while in case of knocking over, help of sinkers or verges is necessary to prevent the old loop from moving down with the needle.

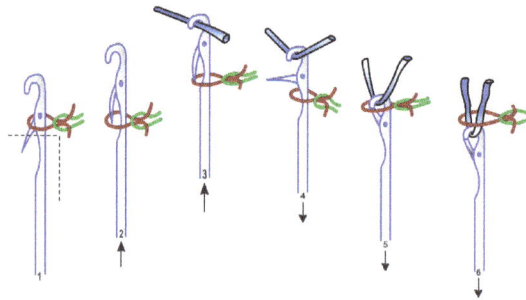

Sequence of loop formation

Course and wale: The horizontal row of loops is called course. The vertical column of loops is called wale. The wales per inch (wpi) and courses per inch (cpi) of knitted fabrics are analogous to ends per inch (epi) and picks per inch (ppi) of woven fabrics. For a fully relaxed knitted fabric, the wpi and cpi values are determined by the loop length. Smaller loop length leads to higher values of wpi and cpi. As a result, the stitch density or loop density which is a product of wpi and cpi also increases with the reduction in loop length. The ratio of cpi and wpi is known as loop shape factor. For fully relaxed single jersey fabric, the loop shape factor is around 1.3.

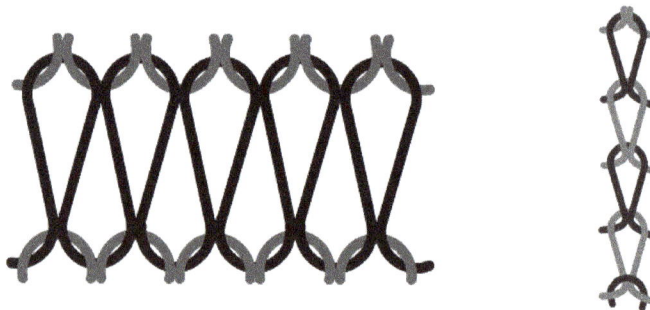

Course and wale

Tightness factor: It indicates the proportion of knitted fabric area covered by the yarn. It is analogous to the cover factor of woven fabric. If the linear density of the yarn is T tex and loop length is L cm, then tightness factor is $\dfrac{\sqrt{T}}{L}$.

Flat bed machines, as the name implies, have one or more beds for carrying the needles. Single bed machines produce plain or single jersey structure whereas double bed machines (V bed) produce rib (1×1, 2×2) and purl structures (figure below). Double bed machines can also be employed to develop single jersey constructions. The needles on the two beds in a double bed machine must be offset so that they do not collide with each other while forming the loops.

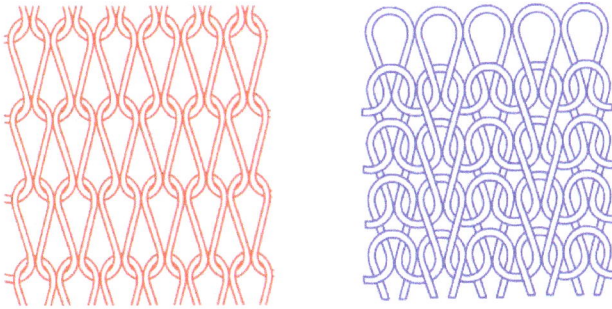

Single jersey and double jersey (rib) structures

In case of single jersey fabrics, all the heads of the loops are either facing the viewer or away from the viewer. In figure (a), all the heads of the loops are hidden from the viewer while the legs are prominently visible. So, it is the technical face side of the fabric. The other side of the fabric is known as technical back.

In case of rib (double jersey) fabrics, in some of the wales, heads of the loops are facing the viewer and vice versa (figure b). So there is no technical face or back in double jersey fabric. Single jersey fabrics tend to curl at the edges. In general, double jersey fabric is thicker and more stretchable in course direction than the single jersey fabric.

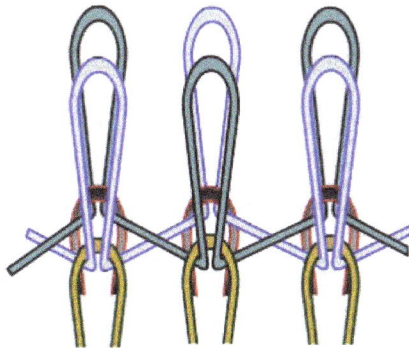

Interlock structure

In circular single jersey knitting machine, only one set of needles are used on the cylinder. However, in circular rib knitting machine, two sets of needles namely cylinder and dial needles are used. They operate perpendicularly to each other. One set of needles (cylinder needles) are arranged on the surface of a grooved cylinder. Generally the cylinder is rotated and needles get the requisite movement from stationary cam jackets. Another set of needles operate in horizontal plane and they are known as dial needles.

Another important double jersey fabric which is made on circular knitting machines is known as interlock which is basically the combination of two rib structures (igure above). The interlocking of two rib structures is responsible for lower stretchability of interlock fabrics as compared to the original rib structure. Interlock fabrics are generally heavy and demonstrate least porosity and highest thermal conductivity among the three knitted structures.

Warp Knitting

Warp knitting is a family of knitting methods in which the yarn zigzags along the length of the fabric, i.e., following adjacent columns ("wales") of knitting, rather than a single row ("course"). For comparison, knitting across the width of the fabric is called weft knitting.

Since warp knitting requires that the number of separate strands of yarn ("ends") equals the number of stitches in a row, warp knitting is almost always done by machine rather than by hand.

History

Credit for the invention is usually given to a mechanic called Josiah Crane in 1775. He likely sold his invention to Richard March who patented (No. 1186) a warp frame in 1778. In the intervening three years March likely had discussed the device with Morris who submitted a similar patent (No.1282) for a twisting machine for making Brussels point lace. These early machines were modifications of the stocking frame with an additional warp beam.

In 1795, the machine was successfully used to make lacey fabrics. Warp frames could be used with any thread, and the warps provided a fixed anchor for the transverse threads. In 1786 Flint invented the point bar which kept the threads at a fixed distance. In 1796, Dawson introduced cams to move the bars, and regulate the twist. Brown and Copstake succeeded in imitating Mechlen net. Lindley invented the bobbin in 1799, and Irving and Skelton the regulator spring. In 1802, Robert Brown of New Radford patented the first twist-frame, a knitter that could produce wide net. Whittaker's frame of 1804 had half its thread mounted on a warp beam and half wound on bobbins mounted on a carriage.

Heathcote's 1808 improvement of Whittaker's frame was essentially a warp knitting frame. The bobbin carrying beam was reduced to the same size as the machine- he called it a bobbinet. Heathcote's second patent, in 1809, was for a bobbinet that could produce wide fabrics- this was the Old Loughborough.

Types

Warp knitting comprises several types of knitted fabrics. All warp-knit fabrics are resistant to runs and relatively easy to sew. Raschel lace— a common type of machine made lace—is a warp knit fabric but using many more guide-bars (12+) than the usual machines which mostly have three or four bars.

Tricot

Tricot is very common in lingerie. The right side of the fabric has fine lengthwise ribs while the reverse has crosswise ribs. The properties of these fabrics include having a soft and 'drapey' texture with some lengthwise stretch and almost no crosswise stretch.

Milanese Knit

Milanese is stronger, more stable, smoother and more expensive than tricot and, hence, is used in better lingerie. These knit fabrics are made from two sets of yarn knitted diagonally, which results in the face fabric having a fine vertical rib and the reverse having a diagonal structure, and results in these fabrics being lightweight, smooth, and run-resistant. Milanese is now virtually obsolete.

Raschel Knit

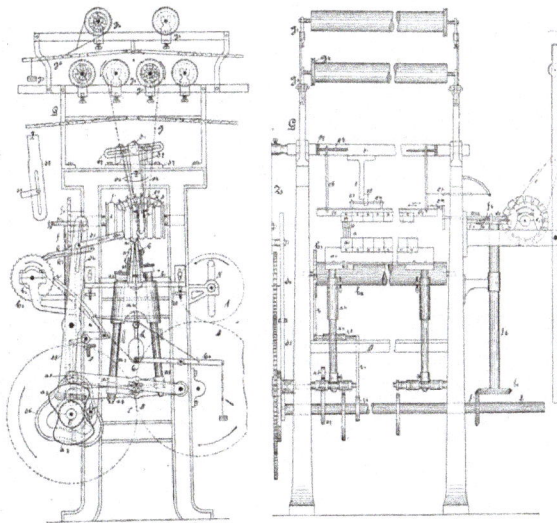

Drawing of an old Raschel machine

Raschel knits do not stretch significantly and are often bulky; consequently, they are often used as an unlined material for coats, jackets, straight skirts and dresses. These fabrics can be made out of conventional or novelty yarns which allows for interesting textures and designs to be created. The qualities of these fabrics range from "dense and compact to open and lofty [and] can be either stable or stretchy, and single-faced or reversible. The largest outlet for the Raschel warp knitting machine is for lace fabric and trimmings. Raschel knitting is also used in outdoors and military fabrics for products such as backpacks. It is used to provide a ventilated mesh next to the user's body (covering padding) or mesh pockets and pouches to facilitate visibility of the contents (MIL-C-8061).

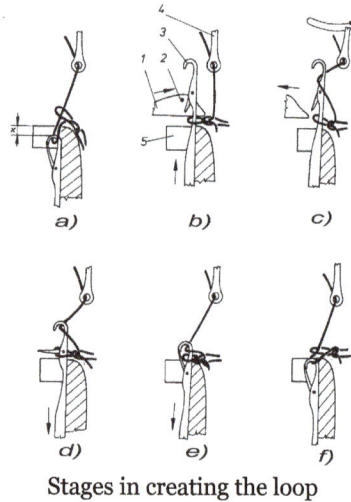

Stages in creating the loop

Golden lace   Lace appliqué   Raschel lace

Stitch-Bonding

- Stitch-Bonding is a special form of warp knitting and is commonly used for the production of composite materials and technical textiles. As a method of production, stitch-bonding is efficient, and is one of the most modern ways to create reinforced textiles and composite materials for industrial use. The advantages of the stitch-bonding process includes its high productivity rate and the scope it offers for functional design of textiles, such as fiber-reinforced plastics. Stitch-bonding involves layers of threads and fabric being joined together with a knitting thread, which creates a layered structure called a multi-ply. This is created through a warp-knitting thread system, which is fixed on the reverse side of the fabric with a sinker loop, and a weft thread layer. A needle with the warp thread passes through the material, which requires the warp and knitting threads to be moving both parallel and perpendicular to the vertical/warp direction of the stitch-bonding machine. Stitch-bonded fabrics are currently being used in such fields as wind energy generation and aviation. Research is currently being conducted into the usage and benefits of stitch-bonded fabrics as a way to reinforce concrete. Fabrics produced with this process offer the potential of using "sensitive fiber materials such as glass and carbon with only little damage, non-crimp fiber orientation and variable distance between threads".

- Extended Stitch-Bonding process (or the extended warp-knitting process): the compound needle that pierces the piles is shifted laterally according to the yarn guides. This then makes it possible for the layers of the stitch-bonded fabric to be arranged freely and be made symmetrical in one working step. This process is advantageous to the characteristics of the composite as the "residual stresses resulting from asymmetric alignment of the layers are avoided, [while] the tensile strength and the impact strength of the composite are improved".

Needle Shift

Needle shift technique is when "Both outer warp layers [are] secured in one procedure by incorporating a shift of the needle bar during the stitching process, creating endless possibilities for the arrangement and patterns in the stitch-bonding process.

Advantages

Producing textiles through the warp knitting process has the following advantages

- higher productivity rates than weaving
- variety of fabric constructions
- large working widths
- low stress rate on the yarn that allows for use of fibers such as glass, aramide and carbon
- the creation of three-dimensional structures that can be knitted on double needle bar raschels

Applications

Warp knitted fabrics have several industrial uses, including producing mosquito netting, tulle fabrics, sports wear, shoe fabric, fabrics for printing and advertising, coating substrates and laminating backgrounds.

Research is also being conducted into the use of warp knitted fabrics for industrial applications (for example, to reinforce concrete), and for the production of biotextiles.

Warp Knitting and Biotextiles

The warp knitting process is also being used to create biotextiles. For example, a warp knitted polyester cardiac support device has been created to attempt to limit the growth of diseased hearts by being installed tightly around the diseased heart. Current research on animals "have confirmed that...the implantation of the device reverses the disease state, which makes this an alternative innovative therapy for patients who have side effects from traditional drug regimes".

Plaited Stitch (Knitting)

In knitting, a plaited stitch is a single knitted stitch that is twisted clockwise (right over left) or counterclockwise (left over right), usually by one half-turn (180°) but sometimes by a full turn (360°) or more.

Methods

Plaited stitches can be produced in several ways. Knitting into the back loop produces a clockwise plaited stitch in the lower stitch being knitted (i.e., the loop that was on the left-hand needle.) The clockwise-plaited stitch is also called a left crossed stitch, since the left strand (i.e., the outgoing strand) of the loop crosses over the right incoming strand. Left-crossed stitches are sometimes called twisted stitches, although the latter term might be confused with similar terms from cable knitting. Conversely, a counter-clockwise plaited stitch can be produced if the yarn is wrapped around the needle in the opposite direction as normal while knitting a stitch. Such a stitch is also called a "right crossed stitch", since the right incoming strand crosses over the left outgoing strand. Here, the plait appears in the upper stitch being knitted, i.e., in the new loop being formed. In the "brute-force" approach, the knitter can produce any sort of plaiting by removing the stitch to be knitted from the left-hand needle, twisting it as desired, then returning it to the left-hand needle and knitting it.

Applications in Knitting

Both clockwise and counterclockwise plaited stitches are often repeated in wales, i.e., in columns of knitting, where they make attractive, subtly different ribbings. Fabrics made with plaited stitches are stiffer than normal and "draw in" sideways, i.e., have a smaller widthwise gauge.

Extra-long, full-turn clockwise plaited stitches can be made by knitting through the back loop and wrapping the yarn twice; this is an attractive stitch when repeated in a row, creating openness and a change in scale that enlivens even simple stockinette or garter stitch.

Plaited stitches are also useful in increases and decreases, both for drawing the fabric together and for covering potential "holes" in the fabric.

As a Method for Correcting Errors

As an aside, knitting through the back loop is a useful technique for untwisting stitches on the left-hand needle that "hang backwards". Such stitches are often produced when a knitted fabric is partially pulled out and some stitches are accidentally put back onto the needle with a backwards twist.

References

- John A. Garraty; Mark C. Carnes (2000). "Chapter Three: America in the British Empire". A Short History of the American Nation (8th ed.). Longman. ISBN 0-321-07098-4

- J.Hausding and C.Cherif, "Improvements in the warp-knitting process and new patterning techniques for stitch-bonded textiles", Journal of the Textile Institute, 2012

- Broudy, Eric (1979). The Book of Looms: A History of the Handloom from Ancient Times to the Present. University Press of New England. pp. 111–112. ISBN 978-0874516494

- Hodder, Ian (2013). "2013 Season Review" (PDF). Çatal Newsletter. pp. 1–2. Archived from the original (PDF) on 2015-04-13. Retrieved 7 February 2014

- Jenkins, D.T., ed. (2003). The Cambridge History of Western Textiles, Volume 1. Cambridge University Press. p. 194. ISBN 978-0521341073

- Scarmeas, N.; Manly, Stern; Tang, Levy (December 26, 2001). "Influence of leisure activity on the incidence of Alzheimer's Disease". Neurology. 57 (12). doi:10.1212/wnl.57.12.2236

- Silva, Marcos (2008). Malharia - Bases De Fundamentação. Universidade Federal do Rio Grande do Norte. p. 2. Retrieved 22 December 2014

- Campbell, Gordon (2006). The Grove Encyclopedia of Decorative Arts, Volume 1. Oxford University Press. ISBN 978-0-19-518948-3

- R.D. Sumanasinghe and M.W. King, "New Trends in Biotextiles-The Challenge of Tissue Engineering", The Journal of Textile and Apparel, Technology and Management, 2003

- Bartholomew Dean 2009 Urarina Society, Cosmology, and History in Peruvian Amazonia, Gainesville: University Press of Florida ISBN 978-0-8130-3378-5

- George Unwin (editor) (1918). "The estate of merchants, 1336-1365: IV - 1355-65". Finance and trade under Edward III: The London lay subsidy of 1332. Institute of Historical Research. Retrieved 18 November 2011

Weaving Designs and Patterns

Design is very important for the aesthetic value of the cloth. Fabric weave design means the pattern that exists between the warp and weft yarns. Bringing changes in the drafting and lifting plan helps in creating a design. Textile weaving design is best understood in confluence with the major topics listed in the following chapter.

Weave Design

Fabric weave design implies the pattern of interlacement between the warp and weft yarns. The design influences the aesthetics as well as the properties of the woven fabrics. The design of woven fabrics is manipulated by changing the following two things.

- Drafting
- Lifting plan

The design is constructed on point paper by using cross (×) and blank. The cross means that the end is passing over the pick. The blank means the end is passing below the pick.

Drafting

Drafting determines the allocation of ends to healds i.e. which end will be controlled by which heald. Generally, drafting is made in such a way that minimum number of healds is required to produce a particular design. This implies that if the interlacement pattern of two ends is identical then they should be controlled by the same heald shaft. In case of drafting, a cross means that the heald is up and a blank means that the heald is down.

Lifting Plan

Lifting plan shows the position of healds (up or down) for different peaks i.e. which heald or healds will be lifted in which pick. It is dependent on the design and the drafting.

Straight Draft

In case of straight draft, a diagonal line is created by the crosses (figure). This implies that, generally, end one is controlled by heald one, end two is controlled by heald two and so on.

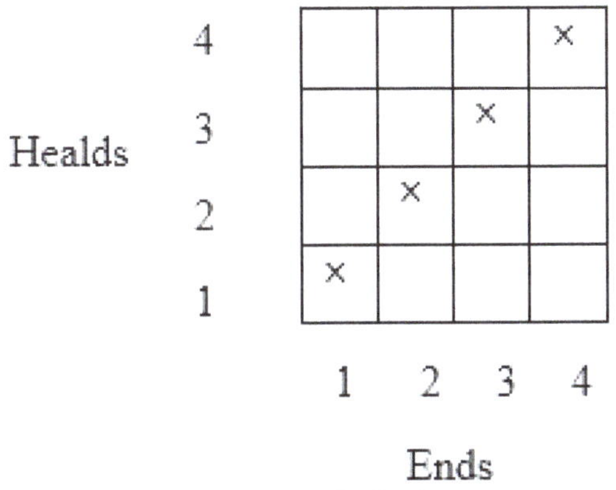

Straight draft

Pointed Draft

In case of pointed draft, a pointed line is created by the crosses (figure below). The repeat of the design contains more than one ends with similar interlacement pattern. For example, in figure, the interlacement pattern is same for ends 1 and 7 and thus they are allocated to one heald (heald number 1). It is also true for ends 2 and 6, 3 and 5, 4 and 8. Therefore, this design which is having eight ends in the repeat requires only four healds. Pointed twill weaves are made using pointed draft.

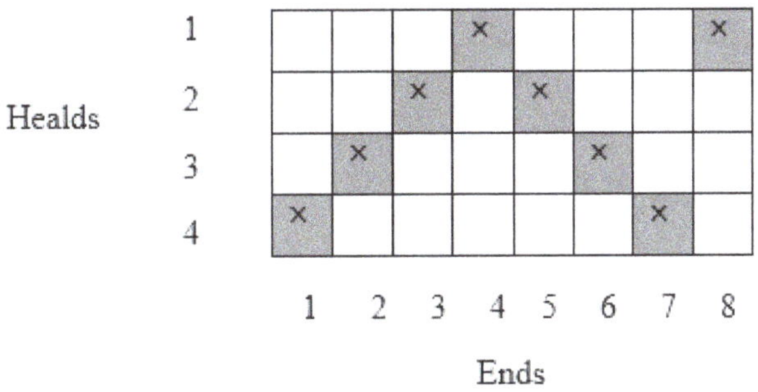

Pointed draft

Skip Draft

In case of skip draft, two or more healds are controlled by a single shedding cam. Plain woven fabrics can be woven with two healds. However, for heavy (high areal density) plain woven fabrics, the number of ends is very high. It often becomes convenient to use four healds for the heavy plain woven fabrics. Therefore, the number of ends controlled by a single heald becomes less as compared to the situation with only two healds. The skip draft for plain woven fabrics with four healds is shown in figure.

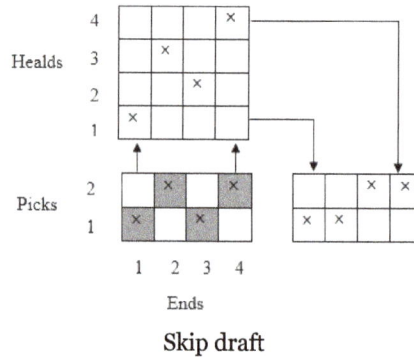

Skip draft

The allocation of healds is shown in the table.

Table: Assigning ends in healds for skip draft

End number	Heald number
1	1
2	3
3	2
4	4

So, the allocation of healds for ends 2 and 3 differs from that of straight draft. Lifting plan shows that for the 1 st pick, healds 1 and 2 are in up position and healds 3 and 4 are in down position. For pick 2, healds 1 and 2 are in down position and healds 3 and 4 are in up position. Therefore, the movement pattern of healds 3 and 4 is just opposite as compared to that of healds 1 and 2. Therefore, healds 1 and 2 can be tied or coupled together with ropes or strings and their shedding operation can be controlled by a single cam. Similarly, shedding operation of healds 3 and 4 can be controlled by another cam. Thus, skip drafting helps to reduce the number of mechanical components (cam, follower, treadle lever etc.) in the loom.

Simple Weaves

The following weaves are the most popular in woven fabrics.

- Plain weave and its derivatives

- Twill weave

- Satin and Sateen weave

Plain Weave

It is the simplest possible and most commonly used weave. The repeat size is 2×2 as depicted in figure a. That implies that the weave repeats on two ends and two picks. It gives maximum number of interlacement in the fabric and therefore the fabric becomes very firm. As the yarns are having maximum possible interlacements, the crimp in the yarns is also higher as compared to other weaves. Figure (b) depicts the interlacement pattern in a plain woven fabric.

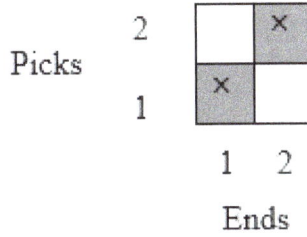

(a) Point paper representation of plain weave

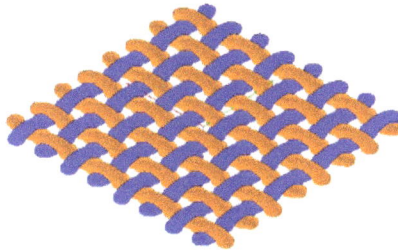

(b) Interlacement pattern of plain weave (warp: orange and weft: blue)

Derivatives of Plain Weave

Warp rib, weft rib and matt (basket) weaves are the derivatives of plain weave. All these designs can be woven with two healds.

Warp Rib

In case of warp rib, two neighbouring picks move in a group as shown in figure. Prominent ribs become visible in the warp direction of the fabrics which is created by the floats of the ends. The picks undergo more number of interlacement than the ends and therefore the crimp in the weft yarns is higher than that of warp yarns.

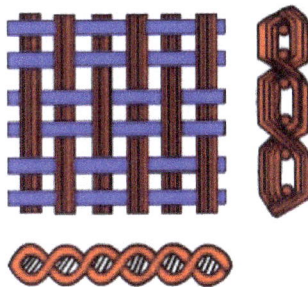

Interlacement pattern of warp rib

Due to the interlacement pattern, warp rib will have more tearing strength in the warp direction as compared to the plain woven fabrics having similar yarns and threads per inch. For warp rib fabrics, two neighbouring picks will resist the tearing force together in a pair resulting in higher tearing strength in warp direction as compared to equivalent (same yarn and same threads per inch) plain woven fabrics.

The design, drafting and lifting plains of warp rib is shown in figure.

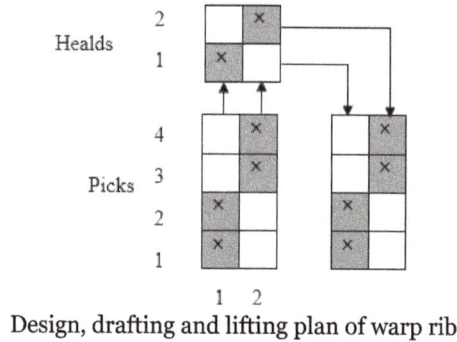

Design, drafting and lifting plan of warp rib

Weft Rib

In case of weft rib, two neighbouring ends move in a group as shown in figure. Prominent ribs become visible in the weft direction of the fabrics which are created by the floats of the picks. The ends undergo more number of interlacement than the picks and therefore the crimp in the warp yarns is higher than that of weft yarns. Weft rib will have more tearing strength in the weft direction as compared to the plain woven fabrics having similar yarns and threads per inch.

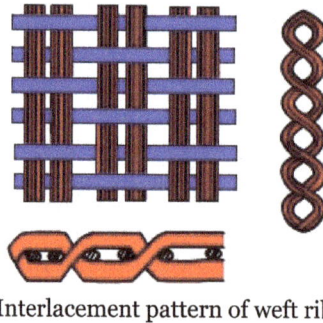

Interlacement pattern of weft rib

The design, drafting and lifting plains of weft rib is shown in figure.

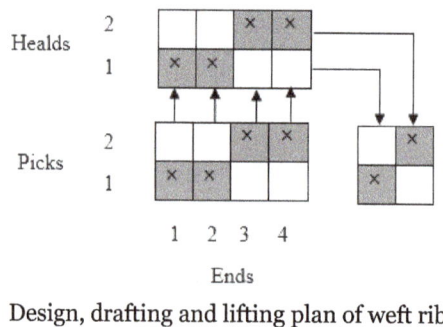

Design, drafting and lifting plan of weft rib

Matt or Basket Weave

In matt weave, multiple ends and picks interlace with each other in a group. The number of interlacement in the fabric is much lower than that of plain weave. In 2×2 matt weave, two ends and two picks form pairs and interlace in the form of plain weave as shown in figure(a). Therefore, the tearing strength of matt woven fabrics is higher in both directions as compared to that of equivalent plain woven fabrics. The design, drafting and lifting plains of 2×2 matt weave is shown in figure (b).

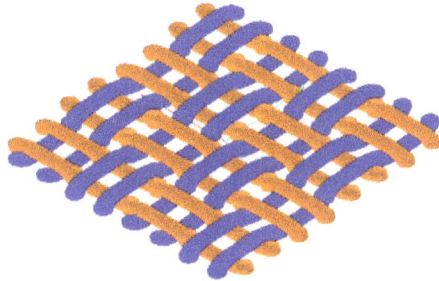

(a) Interlacement pattern of 2×2matt weave (warp: orange and weft: blue)

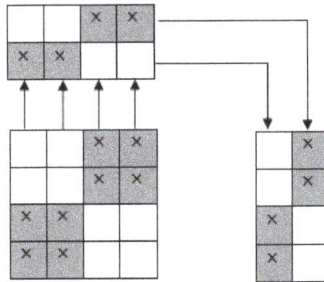

(b) Design, drafting and lifting plan of 2×2matt weave

Twill

A twill weave can be identified by its diagonal lines. This is a 2/2 twill, with two warp threads crossing every two weft threads.

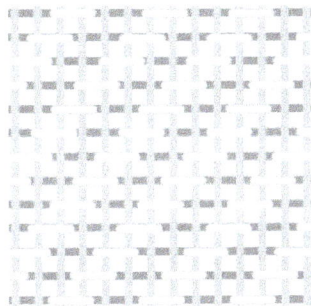

Structure of a $\frac{3}{2}$ twill. The offset at each row forms the diagonal pattern.

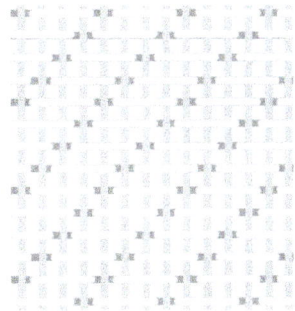

Structure of a $\frac{3}{1}$ twill

Twill is a type of textile weave with a pattern of diagonal parallel ribs (in contrast with a satin and plain weave). This is done by passing the weft thread over one or more warp threads then under two or more warp threads and so on, with a "step," or offset, between rows to create the characteristic diagonal pattern. Because of this structure, twill generally drapes well.

Classification

Twill weaves can be classified from four points of view:

- According to the stepping:
    1. Warp-way: 3/1 warp way twill, etc.
    2. Weft-way: 2/3 weft way twill, etc.
- According to the direction of twill lines on the face of the fabric:
    1. S-Twill or left-hand twill weave: 2/1 S, etc.
    2. Z-Twill or right-hand twill weave: 3/2 Z, etc.
- According to the face yarn (warp or weft):
    1. Warp face twill weave: 4/2 S, etc.
    2. Weft face twill weave: 1/3 Z, etc.
    3. Double face twill weave: 3/3 Z, etc.
- According to the nature of the produced twill line:
    1. Simple twill weave: 1/2 S, 3/1 Z etc.
    2. Expanded twill weave: 4/3 S, 3/2 Z, etc.
    3. Multiple twill weave: 2/3/3/1 S, etc.

Structure

In a twill weave, each weft or filling yarn floats across the warp yarns in a progression of interlacings to the right or left, forming a pattern of distinct diagonal lines. This diagonal pattern is also known as a *wale*. A *float* is the portion of a yarn that crosses over two or more perpendicular yarns.

A twill weave requires three or more harnesses, depending on its complexity and is the second most basic weave that can be made on a fairly simple loom.

Twill weave is often designated as a fraction, such as $\frac{2}{1}$, in which the numerator indicates the number of harnesses that are raised (and thus threads crossed: in this example, two), and the denominator indicates the number of harnesses that are lowered when a filling yarn is inserted (in this example, one). The fraction $\frac{2}{1}$ is read as "two up, one down" (the fraction for plain weave is $\frac{1}{1}$). The minimum number of harnesses needed to produce a twill can be determined by totaling the numbers in the fraction; for the example described, the number of harnesses is three. Twill weave can be identified by its diagonal lines.

Characteristics

Twill fabrics technically have a front and a back side, unlike plain weave, whose two sides are the same. The front side of the twill is called the *technical face* and the back the *technical back*. The technical face side of a twill weave fabric is the side with the most pronounced wale; it is usually more durable and more attractive, is most often used as the fashion side of the fabric, and is the side visible during weaving. If there are warp floats on the technical face (i.e. if the warp crosses over two or more wefts), there will be filling floats (the weft will cross over two or more warps) on the technical back. If the twill wale goes up to the right on one side, it will go up to the left on the other side. Twill fabrics have no "up" and "down" as they are woven.

A twill with ribs in both sides, called *herringbone*

Diamond twill, with weaving edge (left), blue warp, red weft

Sheer fabrics are seldom made with a twill weave. Because a twill surface already has interesting texture and design, printed twills (where a design is printed on the cloth) are much less common than printed plain weaves. When twills are printed, this is typically done on lightweight fabrics.

Soiling and stains are less noticeable on the uneven surface of twills than on a smooth surface, such as plain weaves, and as a result twills are often used for sturdy work clothing and for durable upholstery. Denim, for example, is a twill.

The fewer interlacings in twills as compared to other weaves allow the yarns to move more freely, and therefore they are softer and more pliable, and drape better than plain-weave textiles. Twills also recover from creasing better than plain-weave fabrics do. When there are fewer interlacings, the yarns can be packed closer together to produce high-count fabrics. With higher counts, including high-count twills, the fabric is more durable, and is air- and water-resistant.

Twills can be divided into *even-sided* and *warp-faced*. Even-sided twills include foulard or surah, herringbone, houndstooth, serge, sharkskin, and twill flannel. Warp-faced twills include cavalry twill, chino, covert, denim, drill, fancy twill, gabardine, and lining twill.

Twill Weave

Twill weave is characterised by diagonal line in the fabric which is created by the floats of the ends or picks. The simplest twill weave is two up one down (or one up two down) which repeats on three ends and three picks. Based on the prominence of warp or weft floats, twill weaves are classified as follows.

- Warp faced: 2/1, 3/1, 3/ 2

- Weft faced: 1/2, 1/3, 2/3

- Balanced twill: 2/2, 3/ 3, 2/1 / 1/2

In warp faced twill, the floats of ends predominate over that of picks. In contrast, the floats of picks predominate over that of ends in weft faced twill. In case of balanced twill, the floats of ends and picks are equal. Figure a shows point paper design for a warp faced (2/1) and a balanced (2/2) twill. Figure b and c depict the interlacement pattern for 2/1 and 3/1 twill fabrics respectively. It can be seen from figure c that there are long floats of warp (orange colour) over three consecutive picks visible at the face side of the fabric.

Twill weave has lesser interlacements than the plain weave. Thus the crimp in yarns for twill weave will be lower than that of plain weave. For equivalent fabrics, 3/1 twill will give higher tearing strength than followed by 2/1 twill and plain.

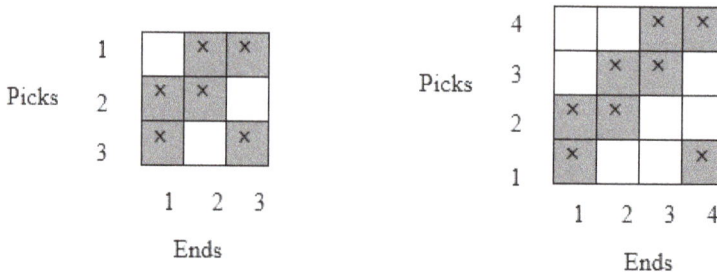

(a) Warp faced (2/1) twill and balanced (2/2) twill

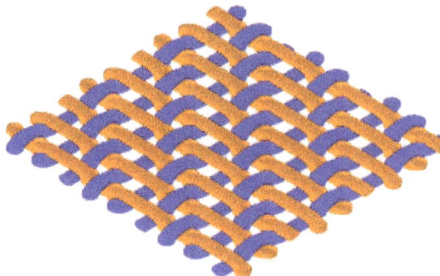

(b) Interlacement pattern in 2/1 twill weave (warp: orange and weft: blue)

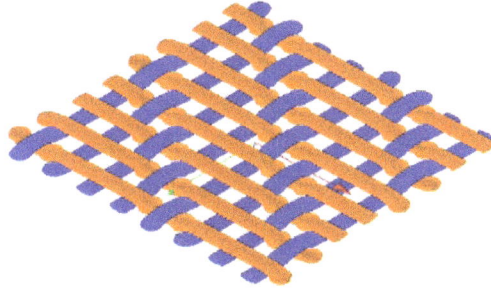

(c) Interlacement pattern in 3/1 twill weave (warp: orange and weft: blue)

Pointed Twill

In pointed twill, there is no continuous line. However, the twill lines change directions at specified intervals and thus create pointed effect on the fabric. The design, drafting and lifting plan of a pointed twill based on basic 2/2 twill weave is shown in the first figure. The 4th end is considered as the mirror line and the design is reversed such that the interlacement pattern for the ends 5, 6 and 7 becomes identical with those of ends 3, 2 and 1, respectively. The interlacement pattern of end 4 and end 8 is same. The pointed twill is woven using the pointed draft as shown in figure. The lifting plan resembles with the left hand side of the design which is true for the pointed draft. Figure depicts the extended view of the same pointed twill.

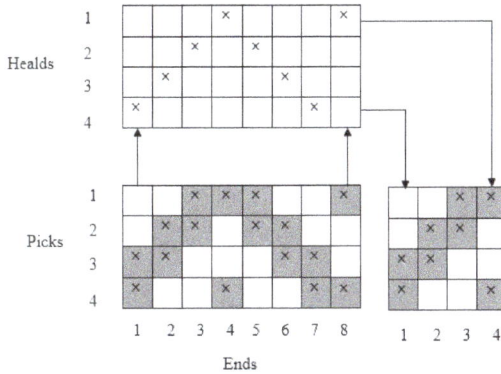

Design, drafting and lifting plan of pointed twill

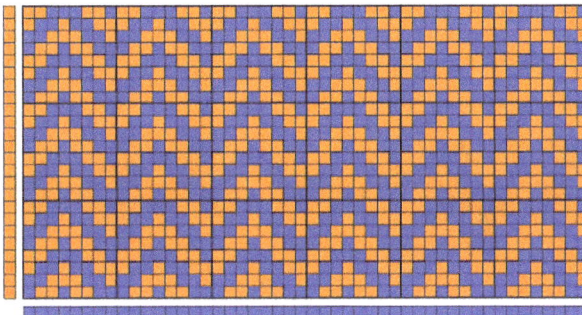

Extended view of pointed twill

Angle of Twill

The angle made by the twill line with the horizontal direction (weft direction) is known as angle of twill or twill angle. From point paper design, it seems that the angle will always be 45°. However, it is dependent on pick spacing, end spacing and move number of the design.

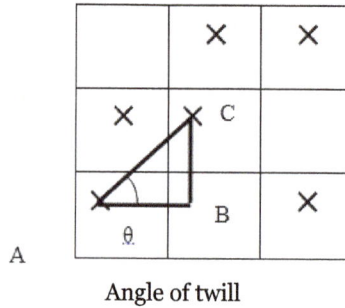

Angle of twill

In figure (Extended view of pointed twill), a 3/1 twill weave has been shown with move number 1 (one step right and one step upward). Here the angle θ (CAB) is the twill angle.

$$Therefore, \theta = \tan^{-1}\left(\frac{BC}{AB}\right) = \tan^{-1}\left(\frac{pick\ spacing}{end\ spacing}\right) = \tan^{-1}\left(\frac{p_2}{p_1}\right)$$

Move number implies the movement of the starting point of the design in horizontal and vertical direction. Generally, for the construction of standard designs, move number 1 is used for both the directions. Therefore, the angle of twill depends on the ratio of pick spacing and end spacing as shown above. However, by using higher move number 2 in the vertical direction, steep twill can be produced which has angle of twill > 45°. On the other hand, by using higher move number in the horizontal direction, reclined twill can be produced which has angle of twill < 45°.

Thus the generalised expression for twill angle is as follows.

$$Twill\ angle(\theta) = \tan^{-1}\left(\frac{p_2}{p_1} \cdot \frac{move\ no.in\ vertical\ direction}{move\ no.in\ horizontal\ direction}\right)$$

Steep twill and reclined twill based on 3/1 twill weave is shown in the first figure. Different twill angles have been depicted in the second figure.

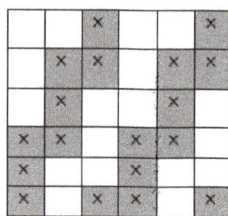

Repeat size 3×6   Repeat size 6×3

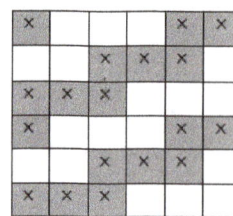

Steep and reclined twill

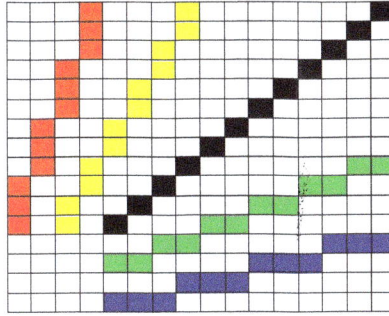

Different twill angles

Satin

Satin is a weave that typically has a glossy surface and a dull back. The satin weave is characterized by four or more fill or weft yarns floating over a warp yarn or vice versa, four warp yarns floating over a single weft yarn. Floats are missed interfacings, where the warp yarn lies on top of the weft in a warp-faced satin and where the weft yarn lies on top of the warp yarns in weft-faced satins. These floats explain the even sheen, as unlike in other weaves, the light reflecting is not scattered as much by the fibres, which have fewer tucks. Satin is usually a warp-faced weaving technique in which warp yarns are "floated" over weft yarns, although there are also weft-faced satins. If a fabric is formed with a satin weave using filament fibres such as silk, nylon, or polyester, the corresponding fabric is termed a *satin*, although some definitions insist that the fabric be made from silk. If the yarns used are short-staple yarns such as cotton, the fabric formed is considered a sateen.

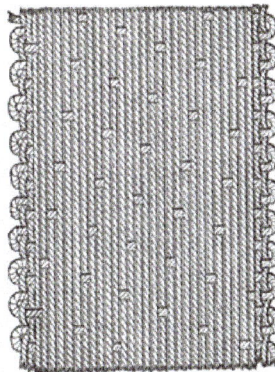

Satin weave. The warp yarns are shown running left and right, weft running top to bottom.

A satin fabric tends to have a high luster due to the high number of floats on the fabric. Because of this it is used in making bed sheets. Many variations can be made of the basic satin weave including a granite weave and a check weave. Satin weaves, twill weaves, and plain weaves are the three basic types of weaving by which the majority of woven products are formed.

Purple satin fabric

Satin is commonly used in apparel: satin baseball jackets, athletic shorts, women's lingerie, nightgowns, blouses, and evening gowns, but also in some men's boxer shorts, briefs, shirts and neckties. It is also used in the production of pointe shoes for use in ballet. Other uses include interior furnishing fabrics, upholstery, and bed sheets .

Satin robe. English, circa 1765

Originally, during the Middle Ages, satin was made of silk; consequently it was very expensive, used only by the upper classes. Satin became famous in Europe during the twelfth century. The name derives its origin from the Chinese port city of Quanzhou, whose name in (medieval) Arabic was Zayton. During the latter part of the Middle Ages, it was a major shipping port of silk, using the maritime Silk Road to reach Europe.

Types of Satin Weaves

Fabrics created from satin weaves are more flexible, with better draping characteristics than plain weaves, allowing them to be formed around compound curves, which is useful in carbon-fiber composites manufacturing. In a satin weave, the fill yarn passes over multiple warp yarns before interlacing under one warp yarn. Common satin weaves are:

- 4-harness satin weave (4HS), also called crowfoot satin, in which the fill yarn passes over three warp yarns and under one warp yarn. It is more pliable than a plain weave.

- 5-harness satin weave (5HS); the fill yarn passes over four warp yarns and then under one warp yarn.

- 8-harness satin weave (8HS), in which the fill yarn passes over seven warp yarns and then under one warp yarn, is the most pliable satin weave and forms most easily around compound curves.

Types of Satin

- Antique satin is a type of satin-back shantung, woven with slubbed or unevenly spun weft yarns.

- Baronet or baronette has a cotton back and a rayon or silk front, similar to georgette.

- Charmeuse is a lightweight, draping satin-weave fabric with a dull reverse.

- Double face(d) satin is woven with a glossy surface on both sides. It is possible for both sides to have a different pattern, albeit using the same colors.

- Duchess(e) satin is a particularly luxurious, heavy, stiff satin.

- Faconne is jacquard woven satin.

- Farmer's satin or Venetian cloth is made from mercerised cotton.

- Gattar is satin made with a silk warp and a cotton weft.

- Messaline is lightweight and loosely woven.

- Polysatin or poly-satin is an abbreviated term for polyester satin.

- Slipper satin is stiff and medium- to heavy-weight fabric.

- Sultan is a worsted fabric with a satin face.

- Surf satin was a 1910s American trademark for a taffeta fabric used for swimsuits.

Sateen

Sateen is a fabric made using a satin weave structure but made with spun yarns instead of filament.

The sheen and softer feel of sateen is produced through the satin weave structure. Warp yarns are floated over weft yarns, for example four over and one under. (In a weft-faced satin or sateen, the weft yarns are floated over the warp yarns). Standard plain weaves use a one-over, one-under structure. The long floats produce a surface that is smooth to the touch and reduces light scattering to increase shine. This weave structure is more susceptible to wear than other weaves.

In modern times cheaper rayon is often substituted for cotton. Better qualities are mercerized to give a higher sheen. Some are only calendered to produce the sheen, but this disappears with washing.

Plain Weave

Plain weave (also called tabby weave, linen weave or taffeta weave) is the most basic of three fundamental types of textile weaves (along with satin weave and twill). It is strong and hard-wearing, used for fashion and furnishing fabrics.

An example of the thread crossing pattern in a plain weave fabric

Structure of plain-woven fabric

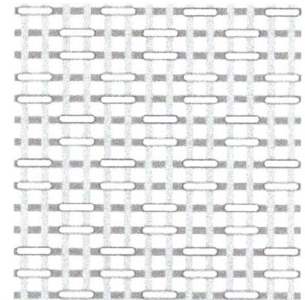

Structure of basketweave fabric

In plain weave, the warp and weft are aligned so they form a simple criss-cross pattern. Each weft thread crosses the warp threads by going over one, then under the next, and so on. The next weft thread goes under the warp threads that its neighbor went over, and vice versa.

- Balanced plain weaves are fabrics in which the warp and weft are made of threads of the same weight (size) and the same number of ends per inch as picks per inch.

- Basketweave is a variation of plain weave in which two or more threads are bundled and then woven as one in the warp or weft, or both.

A balanced plain weave can be identified by its checkerboard-like appearance. It is also known as one-up-one-down weave or over and under pattern.

Examples of fabric with plain weave are chiffon, organza, percale and taffeta.

Designation

According to the 12th-century geographer al-Idrīsī, the city of Almería in Andalusia manufactured imitations of Iraqi and Persian silks called ʿattābī, which David Jacoby identifies as "a taffeta fabric made of silk and cotton (natural fibers) originally produced in Attabiya, a district of Baghdad."

End Uses

Its uses range from heavy and coarse canvas and blankets made of thick yarns to the lightest and finest cambries and muslins made in extremely fine yarns.

Satin and Sateen Weaves

Six-end Regular Sateen

It has been demonstrated earlier that if move number is 1 or n-1 then twill weave is produced. Here, n is the repeat size of the design. If a six-end sateen weave is designed with move numbers of 2, 3 or 4, then the following interlacement pattern will be produced.

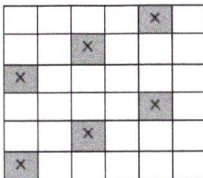

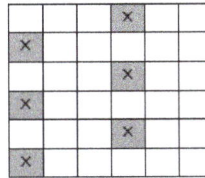

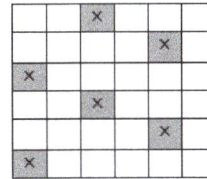

Move number 2   Move number 3   Move number 4

Six-end sateen

In all the three cases, there are certain ends without any interlacement. Therefore, these designs are practically not valid. Therefore, 6 end regular sateen (or satin) weave is not feasible.

Rules for Making Sateen Weave

1. Move number 1 and (n-1) cannot be used as twill weaves are produced.

2. Move number and repeat size of the design should not have any common factor.

It seems from the point paper design that a Satin fabric will become Sateen if the fabric is reversed (turned upside down). However, practically it is not true. Because, satin fabric is warp faced and to make the effect of the warp floates more prominent, following steps are adopted.

•   Use of coarser warp threads than the weft threads

•   Use of higher ends per inch (epi) than the (ppi)

Therefore, even if the fabric is reversed, the effect of weft threads will not be very prominent as the picks are finer and ppi value is lower.

Some Fancy Weaves

Honeycomb

Honeycomb weave shows prominent diamond shapes on the fabrics created by the long floats of ends. Honeycomb weave having a repeat size of 8×8 is shown in the left figure with drafting and lifting plan. The design can be produced with pointed draft and thus the lifting plan resembles the left hand side of the design. The extended view of the Honeycomb weave is shown in the right figure.

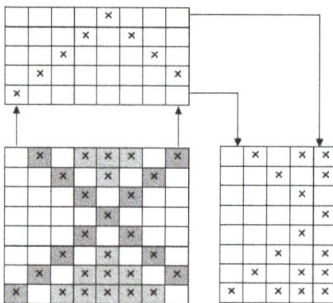

Design, drafting and lifting plan of Honeycomb weave

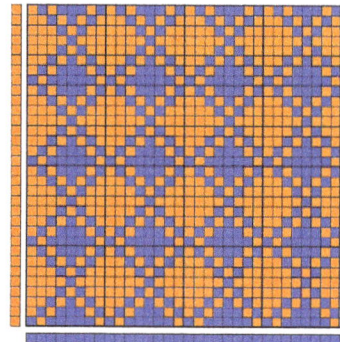

Extended view of Honeycomb weave

Mock Leno

In Mock leno weave, some of the ends have frequent interlacement whereas the other ends have long floats. The fabric shows small holes created by the grouping of threads. A mock leno weave having a repeat size of 10×10 is shown in the left figure along with drafting and lifting plan. Only four healds are needed as the interlacement pattern of ends 1, 3, 5 are same and they are allocated to heald 1. Similarly, the interlacement pattern of ends 2 and 4 are same and they are assigned to heald 2 and so on. Figure right depicts the extended view of the Mock leno weave.

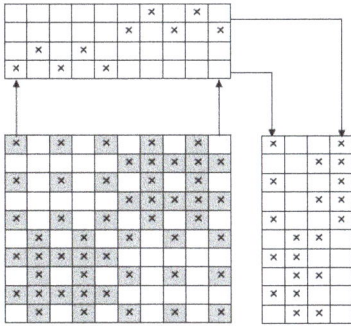

Design, drafting and lifting plan of Mock leno weave

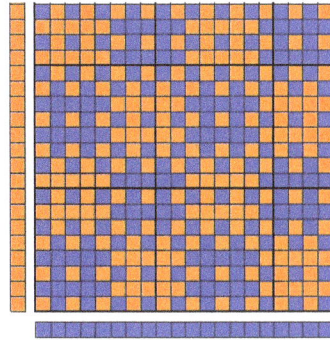

Extended view of the Mock leno weave

Huck-a-back

Huck-a-back design has some similarity with Mock leno. A 10×10 Huck-a-back design is shown in the first figure. If the design is divided in four quadrants, then the top-right and bottom-left corners are having similar interlacement pattern like Mock leno. However, the remaining two quadrants have plain weave like interlacement pattern. Therefore, some of the ends (end number 2, 4, 7 and 9) are having long floats followed by regular interlacements. The design shown below requires four heald shafts. The second figure depicts the extended view of the Huck-a-back weave.

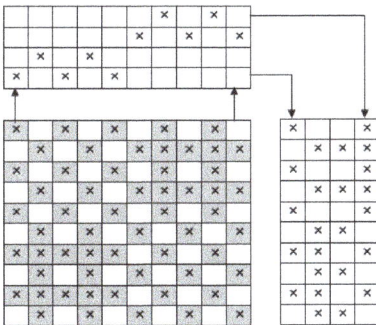

Design, drafting and lifting plan of Huck-a-back weave

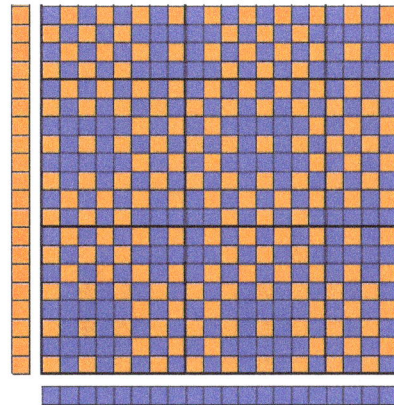

Extended view of the Huck-a-back weave

References

- Emery, Irene (1994). The Primary Structures of Fabrics. Washington, D. C. Thames and Hudson. p. 137. ISBN 978-0-500-28802-3

- Cumming, Valerie; Cunnington, C.W.; Cunnington, P.E. (2010). The dictionary of fashion history. Oxford: Berg. p. 231. ISBN 1847887384

- Tellier, Luc-Normand (2009). Urban world history: an economic and geographical perspective. Presses de l'Université du Québec. p. 221. Retrieved 2010-11-28

- Young, Deborah (2015). Swatch Reference Guide to Fashion Fabrics. Bloomsbury Publishing. p. 117. ISBN 1628926562. Retrieved August 25, 2015

- Tortora, Phyllis G. (2005) Fairchild's Dictionary of Textiles (7th Ed.) New York: Fairchild Publications, p. 490. ISBN 9780870057076

- Kadolph, Sara J., ed.: Textiles, 10th edition, Saddle River, New Jersey, Pearson/Prentice-Hall, 2007, ISBN 0-13-118769-4

Manufacturing Processes of Nonwoven Fabrics

Nonwovens are fabrics that are made from long and staple fibers. The technology used in these fabrics is different from weaving and knitting. Nonwoven fabrics are bonded together by heat, chemical or with solvent treatment. The topics discussed in the chapter are of great importance to broaden the existing knowledge on nonwoven fabrics.

Nonwoven Fabric

Nonwoven fabric is a fabric-like material made from staple fiber (short) and long fibers (continuous long), bonded together by chemical, mechanical, heat or solvent treatment. The term is used in the textile manufacturing industry to denote fabrics, such as felt, which are neither woven nor knitted. Some nonwoven materials lack sufficient strength unless densified or reinforced by a backing. In recent years, nonwovens have become an alternative to polyurethane foam.

Applications

Nonwoven fabrics are broadly defined as sheet or web structures bonded together by entangling fiber or filaments (and by perforating films) mechanically, thermally or chemically. They are flat or tufted porous sheets that are made directly from separate fibers, molten plastic or plastic film. They are not made by weaving or knitting and do not require converting the fibers to yarn. Typically, a certain percentage of recycled fabrics and oil-based materials are used in nonwoven fabrics. The percentage of recycled fabrics vary based upon the strength of material needed for the specific use. In addition, some nonwoven fabrics can be recycled after use, given the proper treatment and facilities. For this reason, some consider nonwovens a more ecological fabric for certain applications, especially in fields and industries where disposable or single use products are important, such as hospitals, schools, nursing homes and luxury accommodations.

Nonwoven fabrics are engineered fabrics that may have a limited life, single-use fabric or a very durable fabric. Nonwoven fabrics provide specific functions such as absorbency, liquid repellence, resilience, stretch, softness, strength, flame retardancy, washability, cushioning, thermal insulation, acoustic insulation, filtration, use as a bacterial barrier and sterility. These properties are often combined to create fabrics suited for specific jobs, while achieving a good balance between product use-life and cost. They can mimic the appearance, texture and strength of a woven fabric and can be as bulky as the thickest paddings. In combination with other materials they provide a spectrum of products with diverse properties, and are used alone or as components of apparel, home furnishings, health care, engineering, industrial and consumer goods.

Non-woven materials are used in numerous applications, including:

Medical

- isolation gowns
- surgical gowns
- surgical drapes and covers
- surgical masks
- surgical scrub suits
- caps
- medical packaging: porosity allows gas sterilization
- gloves
- shoe covers
- bath wipes
- wound dressings
- drug delivery

Filters

- gasoline, oil and air – including HEPA filtration
- water, coffee, tea bags
- pharmaceutical industry
- mineral processing
- liquid cartridge and bag filters
- vacuum bags
- allergen membranes or laminates with non woven layers

Geotextiles

Nonwoven geotextile bags are much more robust than woven bags of the same thickness.

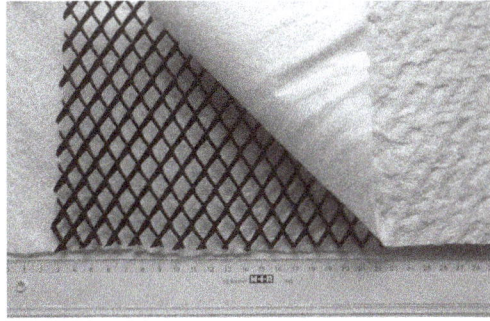

Geocomposite drain consisting of needle-punched nonwoven filter and carrier geotextiles of polypropylene staple fibers each having a mass per area of 200 g/m².

Nonwoven geotextile containers (sand bags) are used for

- soil stabilizers and roadway underlayment
- foundation stabilizers
- erosion control
- canals construction
- drainage systems
- geomembrane protection
- frost protection
- pond and canal water barriers
- sand infiltration barrier for drainage tile
- landfill liners

They are more robust in handling as compared to their woven counterparts, and therefore were often preferred in large-scale erosion protection projects such as those at Amrumbank West; Narrow Neck, Queensland; Kliffende house on Sylt island, and the Eider Barrage. In the last case, only 10 bags out of 48,000 were damaged despite a high installation rate of 700 bags per day.

Other

- diaperstock, feminine hygiene, and other absorbent materials
- carpet backing, primary and secondary
- composites

- o   marine sail laminates
- o   tablecover laminates
- o   chopped strand mat
- backing/stabilizer for machine embroidery
- packaging where porosity is needed
- Shopping bags
- insulation (fiberglass batting)
- acoustic insulation for appliances, automotive components, and wall-paneling
- pillows, cushions, mattress cores, and upholstery padding
- batting in quilts or comforters
- consumer and medical face masks
- mailing envelopes
- tarps, tenting and transportation (lumber, steel) wrapping
- disposable clothing (foot coverings, coveralls)
- weather resistant house wrap
- cleanroom wipes

Manufacturing Processes

Nonwovens are typically manufactured by putting small fibers together in the form of a sheet or web (similar to paper on a paper machine), and then binding them either mechanically (as in the case of felt, by interlocking them with serrated needles such that the inter-fiber friction results in a stronger fabric), with an adhesive, or thermally (by applying binder (in the form of powder, paste, or polymer melt) and melting the binder onto the web by increasing temperature).

Staple Nonwovens

Staple nonwovens are made in 4 steps. Fibers are first spun, cut to a few centimeters length, and put into bales. The staple fibers are then blended, "opened" in a multistep process, dispersed on a conveyor belt, and spread in a uniform web by a wetlaid, airlaid, or carding/crosslapping process. Wetlaid operations typically use 0.25 to 0.75 in (0.64 to 1.91 cm) long fibers, but sometimes longer if the fiber is stiff or thick. Airlaid processing generally uses 0.5 to 4.0 in (1.3 to 10.2 cm) fibers. Carding operations typically use ~1.5" long fibers. Rayon used to be a common fiber in nonwovens, now greatly replaced by polyethylene terephthalate (PET) and polypropylene. Fiberglass is wetlaid into mats for use in roofing and shingles. Synthetic fiber blends are wetlaid along with cellulose for single-use fabrics. Staple nonwovens are bonded either thermally or by using resin. Bonding can be throughout the web by resin saturation or overall thermal bonding or in a distinct pattern via resin printing or thermal spot bonding. Conforming with staple fibers usually refers to a combination with melt blowing, often used in high-end textile insulations.

Melt-blown

Melt-blown nonwovens are produced by extruding melted polymer fibers through a spin net or die consisting of up to 40 holes per inch to form long thin fibers which are stretched and cooled by passing hot air over the fibers as they fall from the die. The resultant web is collected into rolls and subsequently converted to finished products. The extremely fine fibers (typically polypropylene) differ from other extrusions, particularly spun bond, in that they have low intrinsic strength but much smaller size offering key properties. Often melt blown is added to spun bond to form SM or SMS webs, which are strong and offer the intrinsic benefits of fine fibers such as fine filtration, low pressure drop as used in face masks or filters and physical benefits such as acoustic insulation as used in dishwashers. One of the largest users of SM and SMS materials is the disposable diaper and feminine care industry.

Spunlaid Nonwovens

Spunlaid, also called spunbond, nonwovens are made in one continuous process. Fibers are spun and then directly dispersed into a web by deflectors or can be directed with air streams. This technique leads to faster belt speeds, and cheaper costs. Several variants of this concept are available, such as the REICOFIL machinery. PP spunbonds run faster and at lower temperatures than PET spunbonds, mostly due to the difference in melting points

Spunbond has been combined with melt-blown nonwovens, conforming them into a layered product called SMS (spun-melt-spun). Melt-blown nonwovens have extremely fine fiber diameters but are not strong fabrics. SMS fabrics, made completely from PP are water-repellent and fine enough to serve as disposable fabrics. Melt-blown is often used as filter media, being able to capture very fine particles. Spunlaid is bonded by either resin or thermally. Regarding the bonding of Spunlaid, Rieter has launched a new generation of nonwovens called Spunjet. In fact, Spunjet is the bonding of the Spunlaid filaments thanks to the hydroentanglement.

Flashspun

Flashspun fabrics are created by spraying a dissolved resin into a chamber, where the solvent evaporates.

Air-laid Paper

Air-laid paper is a textile-like material categorized as a nonwoven fabric made from wood pulp. Unlike the normal papermaking process, air-laid paper does not use water as the carrying medium for the fiber. Fibers are carried and formed to the structure of paper by air.

Other

Nonwovens can also start with films and fibrillate, serrate or vacuum-form them with patterned holes. Fiberglass nonwovens are of two basic types. Wet laid mat or "glass tissue" use wet-chopped, heavy denier fibers in the 6 to 20 micrometre diameter range. Flame attenuated mats or "batts" use discontinuous fine denier fibers in the 0.1 to 6 range. The latter is similar, though run at much higher temperatures, to melt-blown thermoplastic nonwovens. Wet laid mat is almost always wet resin bonded with a curtain coater, while batts are usually spray bonded with wet or dry resin. An unusual process produces polyethylene fibrils in a Freon-like fluid, forming them into a paper-like product and then calendering them to create Tyvek.

Bonding

Both staple and spunlaid nonwovens would have no mechanical resistance in and of themselves, without the bonding step. Several methods can be used:

- thermal bonding

    o   Use of a heat sealer

    o   using a large oven for curing

    o   calendering through heated rollers (called spunbond when combined with spunlaid webs), calenders can be smooth faced for an overall bond or patterned for a softer, more tear resistant bond.

- hydro-entanglement: mechanical intertwining of fibers by water jets (called spunlace).

- ultrasonic pattern bonding: used in high-loft or fabric insulation/quilts/bedding.

- needlepunching/needlefelting: mechanical intertwining of fibers by needles.

- chemical bonding (wetlaid process): use of binders (such as latex emulsion or solution polymers) to chemically join the fibers. A more expensive route uses binder fibers or powders that soften and melt to hold other non-melting fibers together.

    o   one type of cotton staple nonwoven is treated with sodium hydroxide to shrink bond the mat, the caustic causes the cellulose-based fibers to curl and shrink around one another as the bonding technique

    o   one unusual polyamide(Cerex) is self-bonded with gas-phase acid.

- melt-blown: fiber is bonded as air attenuated fibers intertangle with themselves during simultaneous fiber and web formation.

Disposability

The industry has attempted to define "flushability". They encourage voluntary testing of flushability by producers. They also encourage clear marking of non-flushable products as "No Flush" (rather than fine print on the bottom of products) including creating a "No Flush" logo.

The wastewater industry is encouraging a standard definition (rather than one which varies with each producer) of flushability, including dispersibility, and third-party assessment or verification, such as by NSF International. They believe that products should be both safe for both septic *and* sewer systems (flushable and dispersible, respectively). Orange County Sanitation District has created a campaign, "What 2 Flush", which recommends flushing only the "three P's—pee, poop and [toilet] paper".

Nonwoven Technology

A nonwoven is a sheet of fibres, continuous filaments, or chopped yarns of any nature or origin, that have been formed into a web by any means, and bonded together by any means, with the exception of weaving or knitting. Felts obtained by wet milling are not nonwovens.

Nonwovens are engineered flat structured sheets which are not made by weaving or knitting but by bonding and entangling fibres by means of mechanical, thermal or chemical processes. Nonwoven technology has attracted the attention of the researchers and industrialists as it can manufacture the fabric at a very high production rate bypassing the yarn production stage. The principal end uses of nonwoven materials are in the domain of Technical Textiles such as geotextiles, filtration, wipes, health and hygiene products, surgical gowns, face masks, automotive textiles etc. The two major stages of nonwoven manufacturing are web formation and web bonding. The major nonwoven technologies now available can be listed as shown below:

Web formation	*Web bonding*
Mechanically formed fibre webs (Drylaid)	*Needlepunching*
Aerodynamically formed fibre webs (Drylaid)	*Hydroentanglement*
Hydrodynamically formed fibre webs (Wetlaid)	*Thermal bonding*
Polymer-laid (Spunmelt nonwovens)	*Chemical bonding*

Drylaid, wetlaid and polymer-laid web formation systems have their roots in textile, paper making and polymer extrusion processes.

Needlepunching Technology

Needlepunching is the most common web bonding method. It is the method of consolidation of fibrous webs by the repeated insertion of barbed needles into the web as shown in figure. The needling can be done either from one side or from both (top and bottom) sides of the web. This process consolidates the structure of fibrous web without any binder by interlocking of fibres in the third or 'Z' dimension. Continuous filaments or short staple fibres are initially arranged in the form of a fibrous web in various orientations (random, cross, parallel, or composite). This forms a three-dimensional intermingled structure which fulfils the necessary requirements of geotextiles.

Needlepunched nonwoven geotextiles are extensively used in civil engineering applications including road and railway construction, landfills, land reclamation and slope stabilization. Such applications require geotextiles to perform more than one function including filtration, drainage, and separation. The properties of needle-punched nonwoven depend on parameters like fibre type, web aerial density, needle penetration depth, punch density (number of punches/cm2) and number of needling passages.

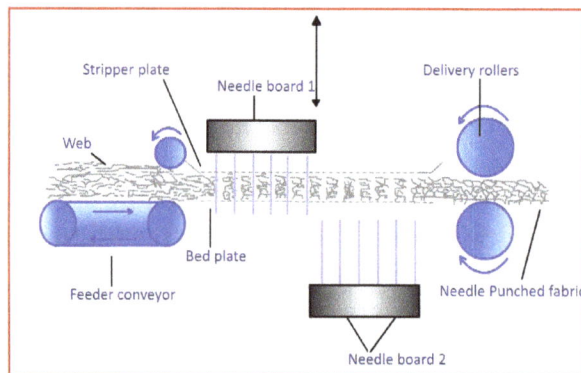

Needlepunching process

Hydroentanglement

Hydroentanglement is a bonding process for wet or dry fibrous webs made by either carding, airlaying or wet-laying, the resulting bonded fabric being a nonwoven. It uses fine, high pressure jets of water which penetrate the web, hit the conveyor belt (or "wire" as in papermaking conveyor) and bounce back causing the fibres to entangle.

Hydroentanglement is sometimes known as spunlacing, this term arising because the early nonwovens were entangled on conveyors with a patterned weave which gave the nonwovens a lacy appearance. It can also be regarded as a two-dimensional equivalent of spinning fibres into yarns prior to weaving. The water pressure has a direct bearing on the strength of the web, and very high pressures not only entangle but can also split fibres into micro- and nano-fibres which give the resulting hydroentangled nonwoven a leatherlike or even silky texture. This type of nonwoven can be as strong and tough as woven fabrics made from the same fibres.

Spunbond Technology

Spunbond and meltblown fabrics belong to the class of polymer-laid nonwovens. In spunbonding process, fluid polymer is converted into finished fabric by a series of continuous operations. Polymer is first extruded into filaments and then the filaments are attenuated. While the filaments are being attenuated, they remain under tension. After attenuation, the tension is released and the filaments are forwarded to a surface where the web is formed. The web is then subjected to the bonding process which can be done by chemical and/or thermal process. A binder may be incorporated in the spinning process or applied subsequently (e.g., a latex). Polypropylene and polyester are commonly used for spun-bonding process. Spun-bended nonwovens have high strength but lower flexibility.

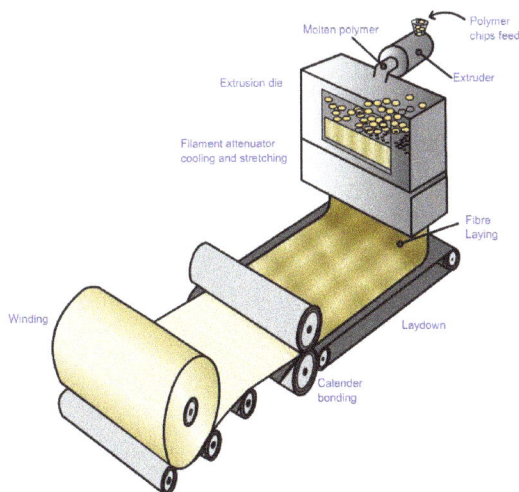

Spunbonding process

Melt Blowing

Melt blowing is a conventional fabrication method of micro- and nanofibers where a polymer melt is extruded through small nozzles surrounded by high speed blowing gas. The randomly deposited fibers form a nonwoven sheet product applicable for filtration, sorbents, apparels and drug delivery systems. The substantial benefits of melt blowing are simplicity, high specific productivity and solvent-free operation.

Melt blowing process

History

During volcanic activity a fibrous material may be drawn by vigorous wind from molten basaltic magma called Pele's hair. The same phenomenon applies for melt blowing of polymers. The first research on melt blowing was a naval attempt in the USA to produce fine filtration materials for radiation measurements on drone aircraft in the 1950s. Later on, Exxon Corporation developed the first industrial process based on the melt blowing principle with high throughput levels.

Polymers

Polymers with thermoplastic behavior are applicable for melt blowing. The main polymer types commonly processed with melt blowing:

- Polypropylene
- Polystyrene
- Polyesters
- Polyurethane
- Polyamides (nylons)
- Polyethylene
- Polycarbonate

Uses

Melt-blown fabrics have generally the same applications as other nonwoven products. The main uses of melt-blown nonwovens and other innovative approaches are as follows.

Filtration

The porous nonwoven melt-blown fabrics can be used in the filtration of gaseous as well as liquid materials. These applications include water treatment, masks, air conditioning filter, etc.

Sorbents

Nonwovens are capable to retain liquids several times of their own weight. For instance, polypropylene nonwovens are ideal to recollect oil contaminations.

Hygiene Products

The high sorption efficiency of melt-blown nonwovens can be exploited in disposable diapers, sanitary napkins and other feminine hygiene products as well.

Apparels

The good thermal insulation properties, the barrier behavior against fluids combined with breathability make melt-blown nonwovens a great choice for apparels even in harsh environments.

Drug Delivery

Melt blowing is also capable to produce drug-loaded fibers for controlled drug delivery. The high throughput rate (extrusion feeding), solvent-free operation accompanied with the increased surface area of the product make melt blowing a promising new formulation technique.

Flashspun Fabric

Flashspun fabric is a nonwoven fabric formed from fine fibrillation of a film by the rapid evaporation of solvent and subsequent bonding during extrusion.

A pressurised solution of, for example, HDPE or polypropylene in a solvent such as fluoroform is heated, pressurised and pumped through a hole into a chamber. When the solution is allowed to expand rapidly through the hole the solvent evaporates to leave a highly oriented non-woven network of filaments.

Air-laid Paper

Air-laid paper is a textile-like material categorized as a nonwoven fabric made from fluff pulp.

Properties

Compared with normal wet-laid paper and tissue, air-laid paper is very bulky, porous and soft. It has good water absorption properties and is much stronger compared with normal tissue.

Main characteristics are:

- Soft, does not scratch.

- Non-linting, no dust, no static.

- Strong, even when wet, can be rinsed and reused.

- Clean, hygienic, can be sterilized.

- Textile-like surface and drape.

- Can be dyed, printed, embossed, coated and made solvent resistant.

Manufacture

Unlike the normal papermaking process, air-laid paper does not use water as the carrying medium for the fibre. Fibres are carried and formed to the structure of paper by air. The air-laid structure is isotropic.

The raw material is long fibered softwood fluff pulp in roll form. The pulp are defibrized in a hammermill. Defibration is the process of freeing the fibres from each other before entering the papermachine. Important parameters for dry defibration are shredding energy and knot content. Normally an air-laid paper consists of about 85% fibre. A binder must be applied as a spray or foam. Alternatively, additional fibres or powders can be added to the pulp which can then be activated and cured by heat.

History

The Danish inventor Karl Krøyer is considered to be the first who commercialized air-formed paper in the early 1980s. Others developed different processes independently at about the same time. A Finnish company United Paper Mills(now UPM-Kymmene Oyj) was one of the companies developing airlaid technology in the 1980s. In the 1980s UPM formed a new company called Walkisoft Oy and also built an airlaid factory to Kotka Finland which started in 1985. Walkisoft built several plants around the world (including the worlds largest airlaid-factory in Mt Holly, NC, USA) in the following 14 years before being sold to Buckeye Technologies Inc. The Walkisoft engineering team, which was responsible for the engineering and R&D of the airlaid machines became known as Buckeye Engineering Finland and from 2002, a private owned Finnish company.

Applications

- Disposable diapers as part of the inner absorbent

- Feminine hygiene

- Industrial wipes

- Personal care products

- Table top

  o napkin

  o table cloth

- Wet wipes

Braiding

Braiding generally produces tubular or narrow fabrics by intertwining three or more strands yarns, threads or filaments. The yarn packages move on serpentine path as shown in figure. In simple machines, half of the packages move in clockwise direction whereas the remaining packages move in anticlockwise direction. Shoelaces and ropes are manufactured using braiding systems. Profiled braided structures are also used as composite performs. The interlacement pattern of braided structure has resemblance with the woven structures. For example, Diamond, Regular and Hercules braids have interlacement patterns similar to those of plain, 2×2 and 3×3 twill weaves respectively.

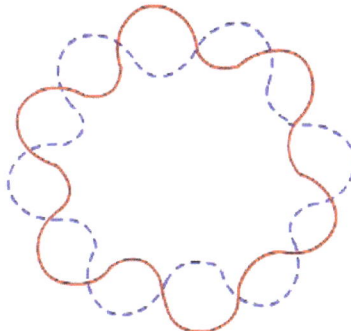

Paths of yarn packages in braiding machine

During the braiding process, the yarn forms a helical path around a mandrel. The braid angle is defined as the angle between the yarn axis and the braid axis. This is similar to the twist angle of a fibre inside the yarn. The braid angle is very important parameter for the braided structure. It can be calculated using the following expression.

$$Braid\ angle = \theta = \tan^{-1}\left(\frac{\omega r}{TUS}\right)$$

where ω is average angular velocity of the package (rad / s)

r is the mandrel radius

TUS is the take – up speed (cm / s)

If the number of yarn careers is K then the braided structure forms K/2 number of parallelograms in the circumferential direction. This is shown in the following figure.

The cover factor for the above braided structure can be calculated using the following expression.

$$Cover\ factor = 1 - \left(1 - \frac{W_y k}{4\pi r \cos\theta}\right)^2$$

where Wy is yarn width

r is the mandrel radius

θ is the braid angle

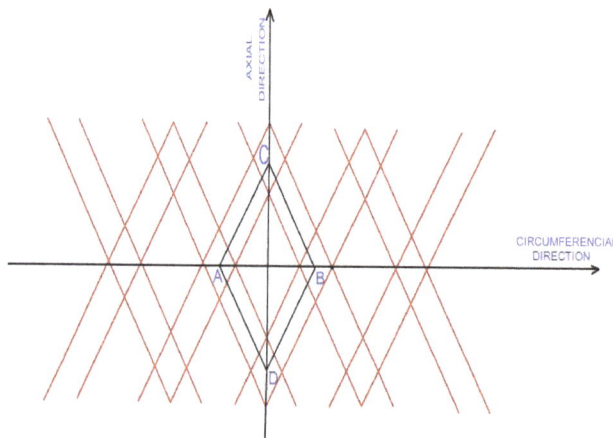

Unit cell of braided structure

Braiding Machine

A braiding machine is device, which interlaces at least three strands of yarns or wires to form a rope reinforced hose, covered power cords, and some types of lace. Materials include natural and synthetic yarns, metal wires, leather tapes and others.

Maypole Braiders

Alternate dancers traveling in opposite directions around a maypole.
Notice how dancers use their arms to raise the ribbons to allow other dancers to pass by.

Maypole braiders work by a circular braiding process. They were well suited to be driven by the steam engines of the industrial revolution and were common by the beginning of the 20th century being easily powered by electric motors. Common types of braiding machines work in much the same way as the process of decorating a maypole. At the start of decorating a maypole an even number of ribbons tied to the top of the pole. A group of people form a ring about the base of the pole and take a ribbon in hand. Half the people then travel clockwise and the other half counter clockwise. When the people pass one another they pass alternately to the right and to the left. This results in a downward forming braid on the pole. As the braid works its way down the pole, the ribbons become shorter and the angle of forming changes as the braid works lower on the pole. On a standard braiding machine, the supply lines are a constant angle and at a constant tension and hence the output braided product is uniform.

Process

1. Fibers are spun into yarn.

2. One or more yarns are twisted together to form a strand.

3. Strands are wound onto bobbins.

4. Bobbins are mounted on carriers.

5. Carriers are mounted onto a braiding machine, where the braiding takes place.

Horn Gear Braider

In a horn gear braider, bobbins of thread pass one another to the left and right on pseudo-sinusoidal tracks, a peg at their bases is driven by a series of so called *horn gears*. A horn gear consists of a notched disk driven by a spur gear below on the same shaft. These gears lie below the track plate that the bobbin carriers ride on, and an even number of gears must be used as there are always an even number of bobbins. The gears must be driven at multiple points on machines with two or more bobbin sets and cross-shafts are used.

A horn gear braiding machine at the Arbetets Museum (Museum of Work) in Norrköping, Sweden

On a vertically oriented machine, the braided thread is taken up above the machine and height and diameter of a guide ring determines the characteristics of the braided product to some degree. On horizontal oriented machines, the braiding track plate and associated bobbins are turned through 90 degrees. This enables large stiff braided cables to be output horizontally so a tall factory building is not required.

Braiding machines, although they have an apparent complex movement of bobbins, are mechanically simple and robust. Modern versions are very reliable and consequently can operate for many hours or even days without attention. This has enabled factories with hundreds of machines to be operated by just a few workers, reducing wages, and making products cheaper and/or profits higher. These modern machines have incorporated electronic controls with automated controls. Although ropes, cords and fishing line are still the core products of most braiding companies there are many other products including webbing, cable shielding and automotive products such as reinforced brake lines.

Some machines use a different system, where the motion of the thread carriers resemble a Teacup Ride.

Square Braider

The illustration shows the base plate of a horn gear braider specifically designed to create a square braid from eight strands of yarn.

The braider has two tracks, one shaded in green and one shaded in red. Four bobbin carriers slide along each track. The carriers shaded in green travel around the green loop in the clockwise direction, and the four carriers shaded in red travel around the red loop in the counter-clockwise direction. As the two tracks cross over, the strands twist around each other, making a braid.

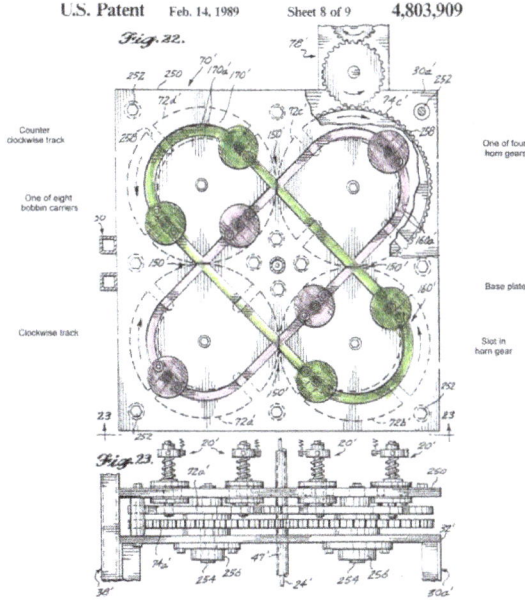

An annotated patent for a square braider

The carriers are pushed by four horn gears in the base plate. Each of the horn gears has a gear and a horn on a common shaft. The gears intermesh with each other, with alternate gears traveling in opposite directions. Each horn has four slots for pushing a bobbin carrier. The bobbin carriers get passed along from one horn to the next as they make their way along a track.

Wardwell Rapid Braider

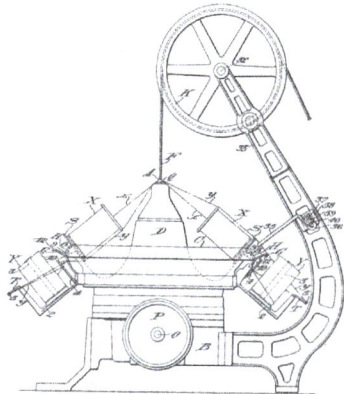

Patent for Wardwell Rapid Braider

The speed of a horn gear braider is limited by the effort needed to force bobbin carriers to follow a serpentine path. In 1922, Simon W. Wardwell solved this problem by moving the strands of yarn, instead of the carriers. The carriers follow a simple circular path, while lever arms driven by a cam guide the strands of yarn to their correct places. Because a lever arm has much less mass than a bobbin and carrier, the machine can run faster. Although these machines run smooth and fast, they must be built for a particular type braid, and are not easily reconfigured.

Track and Column Braider

In a track and column braider, bobbin carriers follow tracks in a two dimensional array of rows and columns, instead of circular paths defined by horn gears.

Automatic High Speed Braiding Machines

Automatic high speed braiding machines are normally equipped with programmable logic controller (PLC), automatic mass adjusting and motor-driven yarn feeders, so that to make the operation simple and easy. Meanwhile, compared to normal braiding machines, high speed braiding machines save costs from labour, electricity and time.

Products from braiding machines are everywhere in life, like shipbuilding, national defense industry, port operations, high pressure and magnet shielding wires and pipes, decorative ropes and belts, shoelace, elastic ropes and belts, cables, etc.

There are non-traditional machines that have a grid array of thread carriers that under computer control can braid over complex shapes.

References

- Horrocks, Richard A.; Annex, Brian H. (2000). Handbook of Technical Textiles. Elsevier. pp. 150–151. ISBN 9781855738966

- Shaumbaugh, R.L. (1988). "A macroscopic view of the melt-blowing process for producing micro-fibers". Ind. Eng. Chem. Res. 27 (12): 2363–2372. doi:10.1021/ie00084a021

- Soares, Carlos (1999). Mechanics of composite materials and structures. Dordrecht Boston, MA: Kluwer Academic Publishers. ISBN 9780792358701. doi:10.1007/978-94-011-4489-6

- M. G. Kamath; Atul Dahiya; Raghavendra R. Hegde (April 2004). "Spunlace (Hydroentanglement)". University of Tennessee's College of Engineering. Retrieved 2013-11-25

- Paulapuro, Hannu (2000). "4". Paper and Board grades. Papermaking Science and Technology. 18. Finland: Fapet Oy. pp. 95–98. ISBN 952-5216-18-7

- Ellison CJ, Phatak A, Giles DW, Macosko CW, Bates FS (2007). "Melt blown nanofibers: Fiber diameter distributions and onset of fiber breakup". Polymer. 48: 3306–3316. doi:10.1016/j.polymer.2007.04.005

- Yordan, Kyosev, (2015-01-01). Braiding technology for textiles. WP, Woodhead Publ./Elsevier. ISBN 9780857091352. OCLC 931672549

- Xiang, P.; Kuznetsov, A. V.; Seyam, A. M. (2008). "A Porous Medium Model of the Hydro entanglement Process". Journal of Porous Media. 11 (1): 35–49. doi:10.1615/JPorMedia.v11.i1.30

- D.V. Rosato and Marlene G. Rosato (2000). Concise Encyclopedia of Plastics. Springer. p. 268. ISBN 978-0-7923-8496-0

- Wulfhorst, Burkhard; Thomas Gries & Dieter Veit (2006). "Braiding Processes and Machines". Textile Technology. Carl Hanser Verlag GmbH & Co. KG: 188–204. doi:10.3139/9783446433472.007

- "Non-Woven In Filtration - A Review | Types Of Filtration | Scope Of Filtration - Textile Mates". Textile Mates. 2017-03-14. Retrieved 2017-03-17

5

An Integrated Study of Winding

Winding as a process can be explained as the transfer of spinning yarn from a package to another. There are different types of winding principles such as spindle-driven winders, pirn winders, yarn tensioning and splicing. The major categories of winding are dealt with great details in the chapter.

Winding

Objectives of Winding

To wrap the forming yarn on a package in a systematic manner or to transfer yarn from one supply package to another in such a way that the latter is adequately compact and usable for the subsequent operations.

To remove the objectionable faults present in original yarns.

Most of the textile winding operations deal with the conversion of ringframe bobbins into cones or cheeses. One ringframe bobbin (cop) typically contains around 100 grams of yarn. If the yarn count is 20 tex, then the length of yarn in the package will be around 5 km. As the warping speed in modern machines is around 1000 m/min, direct use of ringframe bobbins in warping will necessitate package change after every 5 minutes. This will reduce the running efficiency of warping machine. Therefore, ringframe bobbins are converted into bigger cones (mass around 2 kgs or more) or cheeses.

Ringframe bobbins are also not useable as transverse or weft packages because they have empty core which will require bigger size of the shuttle and thereby causing problem in shedding operations. Therefore, for shuttle looms, pirn winding operations are carried out to manufacture weft packages from cones.

Two basic motions are required for effective winding. First, the rotational motion of the package, on which the yarn is being wound, is required. This rotational motion pulls out the yarn from the supply package. Second, the traverse motion is requited so that the entire width of the package is used for winding the yarn. In the absence of the latter, yarns will be wound at the same region by placing one coil over another which is not desirable.

During winding, the yarn can be withdrawn from the supply packages in two ways as depicted in figure.

- Side withdrawal

- Over-end withdrawal

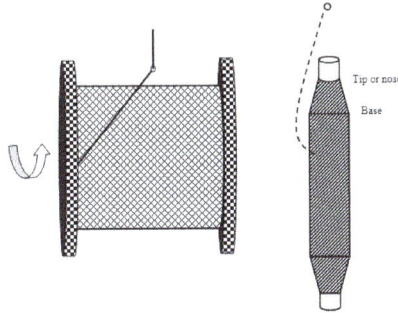

Side withdrawal and over-end withdrawal

Side withdrawal is preferable for flanged packages as the yarn does not touch with the flanges. The package has to rotate during the yarn withdrawal. However, for ringframe bobbins, over end withdrawal is performed by keeping the package in almost upright conditions. As one coil comes out from the ringframe bobbin, one twist is either added or subtracted from parent yarn depending on direction of twist in the yarn.

Types of Wound Packages

There could be three types of wound packages based on the angle at which the yarns are laid on the package.

- Parallel wound package

- Nearly parallel wound package

- Cross wound package

Figure depicts various types of wound packages.

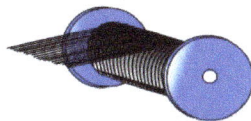

Parallel wound packages

Nearly parallel wound packages

Cross wound packages

Various types of wound packages

In parallel wound package, yarns are laid parallel to each other. This helps to maximize the yarn content in the package. However, parallel wound packages suffer from the problem of stability and layers of coils can collapse specially from the two sides of the package. Therefore, double flanged packages are sometimes used for parallel wound packages.

Example: Weaver's beam, warper's beam.

In nearly parallel wound package, successive coils of yarn are laid with a very nominal angle. The rate of traverse is very slow in this case.

In cross wound package, yarns are laid on the package at considerable angle. As the coils crosses each other very frequently, the package content is lower than that of parallel wound package. However, cross wound package provides very good package stability as the coils often change their direction at the edges of the package.

Example: Cones, Cheeses.

Important Definitions in Winding

Wind: It is the number of revolutions made by the package (i.e. number of coils wound on the package) during the time taken by the yarn guide to make a traverse in one direction (say from left to right) across the package.

Traverse ratio or wind ratio or wind per double traverse: It is the number of revolutions made by the package (i.e. number of coils wound on the package) during the time taken by the yarn guide to make a to and fro traverse. This to and fro traverses of the yarn guide from left to right and back from right to left is known as double traverse.

Traverse ratio= 2× Wind

Angle of wind (θ) : It is the angle made by the yarn with the sides of the package (Figure below). If surface and traverse speeds are V_s and V_t respectively, then

$$\tan \theta = \frac{V_t}{V_s}$$

Coil angle (α) : It is the angle made by the yarn with the axis of the package (Figure below). The coil angle and angle of wind are complementary angles as they add up to 90°.

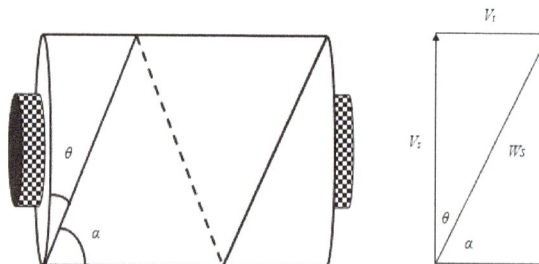

Angle of wind and coil angle

The net winding speed can be obtained by the resultant vector of surface speed (V_s) and traverse speed (V_t).

$$Winding\ speed = \sqrt{Surface\,speed^2 + Traverse\,speed^2}$$
$$= \sqrt{V_s^2 + V_t^2}$$

Winding Machine

Figure depicts the simplified view of a winding machine. It has three main zones.

- Uniwinding zone
- Yarn tensioning and clearing zone
- Winding zone

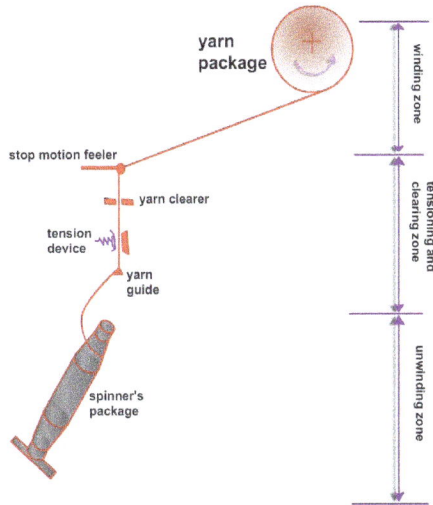

Zones of winding machine

In the unwinding zone, yarns are unwound from the supply package which is ringframe bobbin in most of the cases. Yarn balloon is formed due to the high speed unwinding of yarn from the supply package. Unwinding tension varies continuously as the winding point shifts from tip to base oof a ringframe bobbin and vice versa. Besides, the height of the balloon also increases as the supply package becomes empty.

In the second zone, tensions are applied on the yarns by using tensioners so that yarns are wound on the package with proper compactness. The objectionable yarn faults as well as other contaminants (coloured and foreign fibres) are also removed by using optical or capacitance based yarn clearer.

In the third and final zone, yarns are wound on the package by means of rotational motion of the package and traverse motion of the yarn guide. Based on the operating systems employed in the winding zone, two major winding principles have evolved.

Drum-driven Winders

Primarily there are two types of winding principles as given below.

- Drum-driven or random winders

- Spindle-driven or precision winders

In drum-driven wider, the package is driven by a cylinder by surface or frictional contact as shown in figure. Traverse of yarn is given either by the grooves cut on the drum as shown at the bottom of the second figure or by a reciprocating guide. In case of grooved drum, the drum performs the dual functions of rotating the package by surface contact and performing the traverse (figure a). However, when plain drums are used, it just rotates the package and traverse is performed by reciprocating guide (figure b).

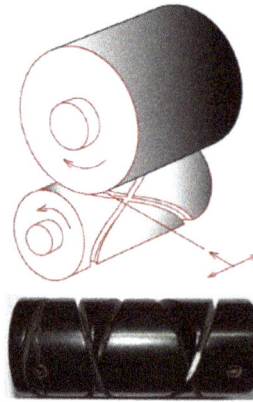

Drum-driven winder and grooved drum

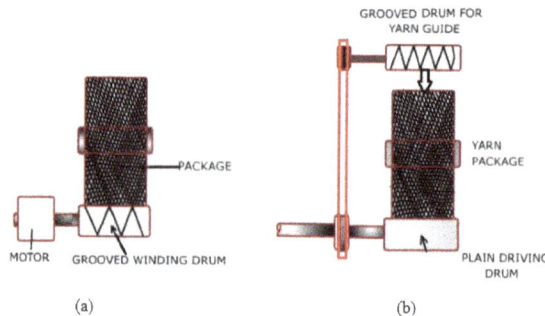

Types of drum-driven winder (a: grooved drum, b: plain drum)

In spindle-driven winder, the package is mounted on a spindle which is driven positively by a gear system. If the r.p.m. of the spindle is constant then the surface speed of the package will increase with the increase in package diameter. Therefore, principle wise there could be two types of spindle-driven winders (following figure).

- Constant r.p.m. spindle winders

- Variable r.p.m. spindle winders

In case of the latter, the spindle r.p.m. is reduced with the increase in package diameter in such a manner that the winding speed remains constant.

Spindle-driven winders are also known as precision winders as a precise ratio is maintained between the r.p.m. of spindle and r.p.m. of traversing mechanism. This leads to maintaining a precise distance between adjacent coils, termed Gain of Wind. The precision winders thus permit precise laying of coils on package and hence its name. Precision winders ensure a constant value of traverse ratio during package building. Precision winders are preferred for winding delicate yarns as the package is not rotated by the surface contact and therefore the possibility of yarn damage due to abrasion is lower as compacted to that of surface driven winders.

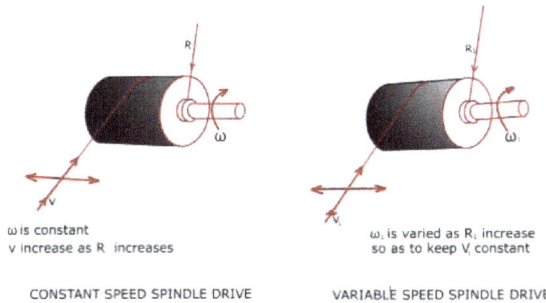

CONSTANT SPEED SPINDLE DRIVE   VARIABLE SPEED SPINDLE DRIVE

Spindle-driven winders

Drum-driven Winders

Let us consider that the diameters of the driving drum and package are D and d respectively. The r.p.m. of drum and package are N and n respectively. D is constant whereas d increases with time due to the building of the package (formation of layers of coils). If there is no slippage between the drum and the package then the surface speed of drum and package will always be same. So, N×D = n×d. The drum r.p.m. N is constant as it getting drive from gear systems and thus n reduces with time.

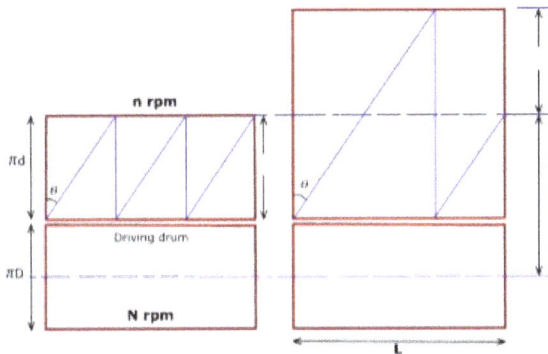

Principles of drum-driven winder

As the drum r.p.m. N is constant, for drum-driven winder, traverse speed and surface speed are also constant. Therefore it gives constant angle of wind and winding speed.

Let, L is the length of the drum and package.

Distance covered in one double traverese = 2L

Number of revolution required for the drum for double traverse = S

N revolution of drum takes 1 minute

S revolution of drum will take $\dfrac{S}{N}$ minute

S revolution of drum is equivalent to one double traverse

So, time for one double traverse= $\dfrac{S}{N}$ min

$$Traverse\ speed = \dfrac{Distance\ covered\ in\ one\ double\ traverse}{Time\ for\ one\ double\ traverse} = \dfrac{2L}{\dfrac{S}{N}} = \dfrac{2LN}{S}$$

$$\tan\theta = \dfrac{V_t}{V_s} = \dfrac{\dfrac{2LN}{S}}{\pi DN} = \dfrac{2L}{\pi DS} = constant\left(as\ L, D\ and\ S\ are\ constant\ for\ a\ given\ drum\right)$$

So, in drum-driven winder, angle of wind remains constant with the increase in package diameter.

Now, if the package r.p.m. is n, same (n) number of coils (wind) will be laid on the package in every minute. Because, one revolution of package creates one coil or wind on the package.

$$So, traverse\ ratio = wind\ /\ double\ traverse$$

$$= \dfrac{wind\ /\ min}{double\ traverse\ /\ min} = \dfrac{n}{\left(\dfrac{N}{S}\right)} = S.\dfrac{n}{N} = S.\dfrac{D}{d}$$

So, in drum-driven winder, traverse ratio reduces with the increase in package diameter.

As S and D are constant for a given drum, the traverse ratio decreases as the package diameter (d) increases. This leads to a 'patterning' problem in case of drum-driven winder.

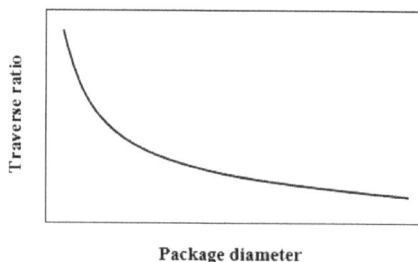

Package diameter
Package diameter vs traverse/ wind ratio

$$Winding\ speed = \sqrt{Surface\ speed^2 + Traverse\ speed^2}$$

$$= \sqrt{(\pi DN)^2 + \left(\frac{2LN}{S}\right)^2}$$

$$= \sqrt{(\pi dn)^2 + \left(\frac{2LN}{S}\right)^2} \quad (no\ slippage\ between\ drum\ and\ package)$$

It is evident from the above expression that the winding speed remains constant during package building in case of drum-driven winder.

Pattern (Sewing)

In sewing and fashion design, a pattern is the template from which the parts of a garment are traced onto fabric before being cut out and assembled. Patterns are usually made of paper, and are sometimes made of sturdier materials like paperboard or cardboard if they need to be more robust to withstand repeated use. The process of making or cutting patterns is sometimes condensed to the one-word Patternmaking but it can also be written pattern making or pattern cutting.

Cutting patterns in a sewing class.

A sloper pattern (home sewing) or block pattern (industrial production) is a custom-fitted, basic pattern from which patterns for many different styles can be developed. The process of changing the size of a finished pattern is called grading.

Several companies specialize in selling pre-graded patterns directly to consumers who will sew the patterns at home. Commercial clothing manufacturers make their own patterns in-house as part of their design and production process, usually employing at least one specialized patternmaker. In bespoke clothing, slopers and patterns must be developed for each client, while for commercial production, patterns will be made to fit several standard body sizes.

Pattern Making

A patternmaker typically employs one of two methods to create a pattern.

The flat-pattern method is where the entire pattern is drafted on a flat surface from measurements, using rulers, curves and straight-edges. A pattern maker would also use various tools such as a notcher, drill and awl to mark the pattern. Usually, flat patterning begins with the creation of a sloper or block pattern, a simple, fitted garment made to the wearer's measurements. For women, this will usually be a jewel-neck bodice and narrow skirt, and for men an upper sloper and a pants sloper. The final sloper pattern is usually made of cardboard or paperboard, without seam allowances or style details (thicker paper or cardboard allows repeated tracing and pattern development from the original sloper). Once the shape of the sloper has been refined by making a series of mock-up garments called *toiles* (UK) or *muslins* (US), the final sloper can be used in turn to create patterns for many styles of garments with varying necklines, sleeves, dart placements, and so on. The flat pattern drafting method is the most commonly used method in menswear; menswear rarely involves draping. There are many pattern making system available, but M. Muller & Sohn is widely used for its accuracy and fit for different body figure.

The draping method involves creating a muslin mock-up pattern by pinning fabric directly on a form, then transferring the muslin outline and markings onto a paper pattern or using the muslin as the pattern itself.

Pattern Digitizing

After a paper/fabric pattern is completed, very often pattern-makers digitize their patterns for archiving and vendor communication purposes. The previous standard for digitizing was the digitizing tablet. Nowadays, automatic option such as scanner and cameras systems are available.

Fitting Patterns

Fitting a muslin on a dress form. This dress form is adjustable to match the sewer's measurements, and the muslin has been fit around the form accordingly, by taking it in two inches at the back, a smaller fit than the original pattern.

Although a sewer may choose to use a standard size that has been pre-graded on a purchased pattern, they may decide to tailor a pattern to better fit the garment wearer. There are several ways this can be done.

Creating a sewer's muslin (also called toile using calico), similar to a garment template, is one method of fitting. Muslin material is inexpensive and is easy to work with when making quick adjustments by pinning the fabric around the wearer or a dress form. The sewer cuts muslin pieces using the same method that they will use for the actual garment, according to a pattern. The muslin pieces are then fit together and darts and other adjustments are made. This provides the sewer with measurements to use as a guideline for marking the pattern pieces and cutting the fabric for the finished garment.

Pattern Grading

Pattern grading is the process of shrinking or enlarging a finished pattern to accommodate it to people of different sizes. Grading rules determine how patterns increase or decrease to create different sizes. Fabric type also influences pattern grading standards. The cost of pattern grading is incomplete without considering marker making.

Standard Pattern Symbols

Sewing patterns typically include standard symbols and marks that guide the cutter and/or sewer in cutting and assembling the pieces of the pattern. Patterns may use:

- Notches, to indicate:
    - Seam allowances. (not all patterns include allowances)
    - Centerlines and other lines important to the fit like the waistline, hip, breast, shoulder tip, etc.
    - Zipper placement
    - Fold point for folded hems and facings
    - Matched points, especially for long or curving seams or seams with ease. For example, the Armscye will usually be notched at the point where ease should begin to be added to the sleeve cap. There is usually no ease through the underarm.
- Circular holes, perhaps made by an awl or circular punch, to indicate:
    - A dart apex
    - Corners, as they are stitched, i.e. without seam allowances
    - Pocket placement, or the placement of other details like trimming
    - Buttonholes and buttons

- A long arrow, drawn on top of the pattern, to indicate:

    o Grainline, or how the pattern should be aligned with the fabric. The arrow is meant to be aligned parallel to the straight grain of the fabric. A long arrow with arrowheads at both ends indicates that either of two orientations is possible. An arrow with one head probably indicates that the fabric has a direction to it which needs to be considered, such as a pattern which should face up when the wearer is standing.

- Double lines indicating where the pattern may be lengthened or shortened for a different fit

- Dot, triangle, or square symbols, to provide "match points" for adjoining pattern pieces, similar to putting puzzle pieces together

Many patterns will also have full outlines for some features, like for a patch pocket, making it easier to visualize how things go together.

Patterns for Commercial Clothing Manufacture

Marker making by computer.

The making of industrial patterns begins with an existing block pattern that most closely resembles the designer's vision. Patterns are cut of oak tag (manila folder) paper, punched with a hole and stored by hanging with a special hook. The pattern is first checked for accuracy, then it is cut out of sample fabrics and the resulting garment is fit tested. Once the pattern meets the designer's approval, a small production run of selling samples are made and the style is presented to buyers in wholesale markets. If the style has demonstrated sales potential, the pattern is graded for sizes, usually by computer with an apparel industry specific CAD program. Following grading, the pattern must be vetted; the accuracy of each size and the direct comparison in laying seam lines is done. After these steps have been followed and any errors corrected, the pattern is approved for production. When the manufacturing company is ready to manufacture the style, all of the sizes of each given pattern piece are arranged into a marker, usually by computer. A marker is an arrangement of all of the pattern pieces over the area of the fabric to be cut that minimizes fabric waste while maintaining the desired grain-lines. It's sort of like a pattern of patterns from which all pieces will be cut. The marker is then laid on top of the layers of fabric and cut. Commercial markers often include multiple sets of patterns for popular sizes. For example: one set of size Small, two sets of size Medium and one set of size Large. Once the style has been sold and delivered to stores – and if it proves to be quite popular – the pattern of this style will itself become a block, with subsequent generations of patterns developed from it.

Retail Patterns

Home sewing patterns are generally printed on tissue paper and sold in packets containing sewing instructions and suggestions for fabric and trim. They are also available over the Internet as downloadable files. Home sewers can print the patterns at home or take the electronic file to a business that does copying and printing. Major pattern companies such as *Burda Style* and independent designers such as Amy Butler distribute sewing patterns as electronic files as an alternative to, or in place of, pre-printed packets. Modern patterns are available in a wide range of prices, sizes, styles, and sewing skill levels, to meet the needs of consumers.

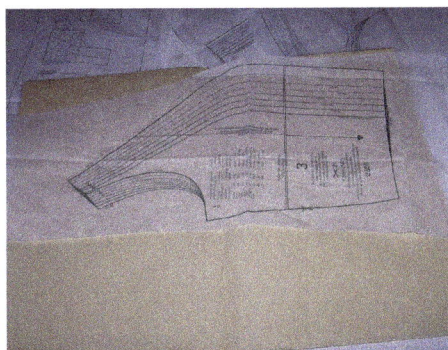

Home tissue paper sewing pattern

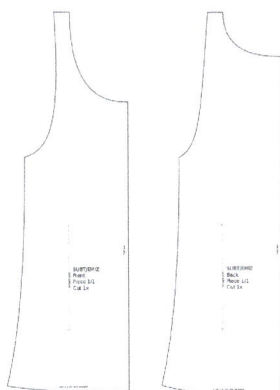

Digital home sewing pattern

The majority of modern-day home sewing patterns contain multiple sizes in one pattern. Once a pattern is removed from a package, you can either cut the pattern based on the size you will be making or you can preserve the pattern by tracing it. The pattern is traced onto fabric using one of several methods. In one method, tracing paper with transferable ink on one side is placed between the pattern and the fabric. A tracing wheel is moved over the pattern outlines, transferring the markings onto the fabric with ink that is removable by erasing or washing. In another method, tracing paper is laid directly over a purchased pattern, and the pieces are traced. The pieces are cut, then the tracing paper is pinned and/or basted to the fabric. The fabric can then be cut to match the outlines on the tracing paper. Vintage patterns may come with small holes pre-punched into the pattern paper. These are for creating tailor's tacks, a type of basting where thread is sewn into the fabric in short lengths to serve as a guideline for cutting and assembling fabric pieces.

Besides illustrating the finished garment, pattern envelopes typically include charts for sizing, the number of pieces included in a pattern, and suggested fabrics and necessary sewing notions and supplies.

Ebenezer Butterick invented the commercially produced graded home sewing pattern in 1863 (based on grading systems used by Victorian tailors), originally selling hand-drawn patterns for men's and boys' clothing. In 1866, Butterick added patterns for women's clothing, which remains the heart of the home sewing pattern market today.

Patterning

Path of Yarn On Cheese

The first figure shows the path of yarn, for one traverse, on a cheese having diameter d and height L. It is seen that the yarn has covered half of the package periphery when it has traversed half of the package height (L/2). This has been depicted with solid line in the above figure. The yarn has made one complete coil (wind) when it has traversed the full height of the package. The path of yarn in the second half of traverse has been shown with a broken line as it cannot be seen from the front view of the package. The situation can be visualised better if the package is cut along its axis and opened as shown in the second figure.

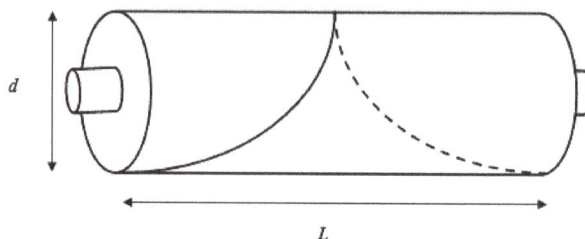

Path of yarn on cheese package

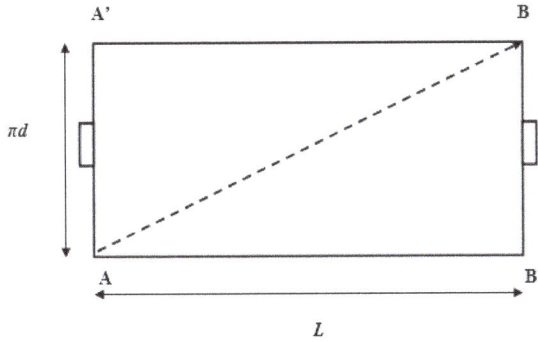

Path of yarn on cheese after cutting

The yarn has started from A and moved to B. So it has made one coil and completed one traverse. If the process is continued then the yarn will make two coils in one double traverse (i.e. traverse ratio = 2). The path of yarn in one double traverse is shown in the figure below.

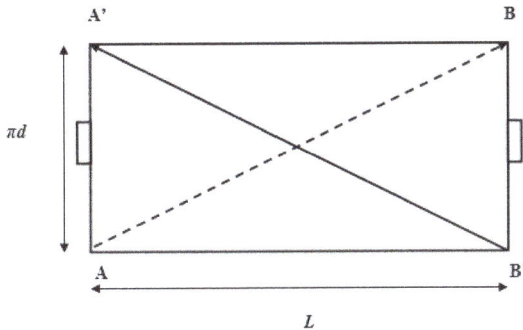

Path of yarn for traverse ratio value 2

So, the path of the yarn in one double traverse can be simply written as ABB'A'. It must be kept in mind that on the package, A and A' are same points and similarly B and B' are same points. It is understandable that after one double traverse the yarn has returned to the same position on the package from where it started its journey. So, during the next double traverse, the yarn will be wound following the same path on the package. This will lead to the formation of ribbon on the package which is known as patterning problem.

Drawing the Path of Yarn on Cheese

For any value of traverse ratio (wind per double traverse) the path of yarn on the cheese can be drawn and analysed by following steps.

- Traverse ration (wind per double traverse) = x (integer)

- Wind per traverse = x/2

- Traverse per wind = 2/x

- Divide the opened package in two equal parts (as the numerator is 2) in the vertical direction and x number of equal parts in the horizontal directions. This will create some smaller rectangles within the opened package.

- Draw the diagonals for the small rectangles.

- When one coil is complete, shift the winding point from upper parallel line to lower parallel line and vice-versa.

- Reverse the direction of traverse when the one traverse is complete.

If traverse ratio is 3 then the value of wind per traverse will be 3/2. So, the value of traverse per wind will be 2/3. As per the steps stated above, the opened package has to be divided into two parts in the vertical direction and three parts in horizontal direction. This is shown in the figure below.

The yarn will move from A to B to complete one coil. B and B' are the same point on the package. The yarn will then move from B' to C to complete one traverse. It can be seen from the figure below that while making one traverse, the yarn completes one and half coils. Then the direction of traverse will change and now the yarn will move from C to B (right to left). Finally the yarn will complete the double traverse by moving from B' to A'. Here, the yarn comes back to the starting position (A) after only one double traverse.

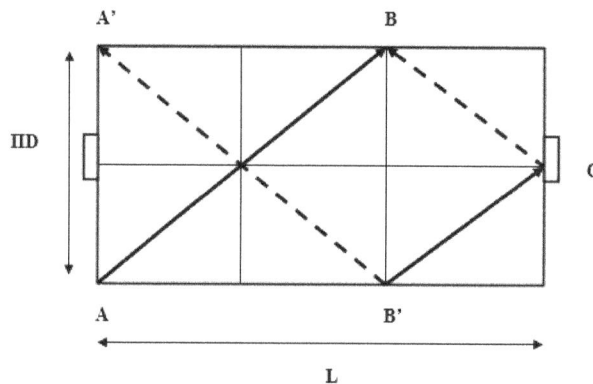

Path of yarn for traverse ratio value 1.5

The following figure depicts the path of yarn for traverse ratio of 7/2. It should be noted that yarn comes back to the starting position after two complete double traverses. By extending this analogy, it can be inferred that if the traverse ratio value of 15/4, then the yarn will come back to the starting position after four double traverses. However, as the number of double traverse increases, number of coils formed on the package also increases which leads to the increase in package diameter. So, the yarn actually comes back to a different point precluding the possibility of patterning. Therefore, patterning is prevalent when the traverse ratio is integer or having values like 1.5 or 2.5 etc.

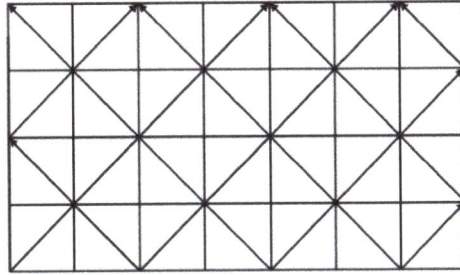

Path of yarn for traverse ratio value 3.5

Spindle-driven Winders

The schematic diagram of a spindle-driven winder is shown in figure. The spindle carrying the package is rotating at n r.p.m. A and B (representing the respective tooth number) are the two gears responsible for transmitting the rotational motion from the spindle to traverse mechanism.

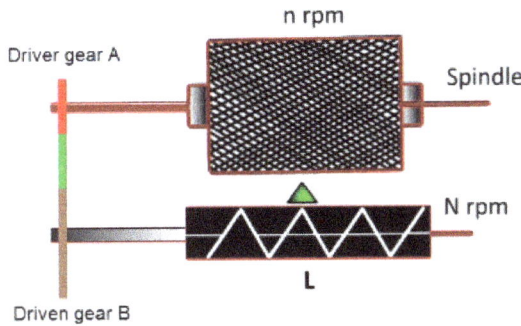

Drive of spindle-driven winder

If these gears (A and B) are not changed then the ratio of spindle speed (r.p.m.) and traverse speed (number of traverse/ min) remains same and therefore the value of traverse ratio remains constant. However, the angle of wind changes during the package building which can be understood from the following expressions.

Let R is the number of double traverse made by the traversing device per minute.

$$\tan\theta = \frac{Travrese\,speed}{Surface\,speed}$$

$$= \frac{V_t}{V_s} = \frac{2LR}{\pi dn}$$

$$Now, r.p.m. \, of \, the \, \text{traversing drum} = N' = n.\frac{A}{B}$$

If the traverse is given by a groove drum which requires S revolutions for one double traverse then,

$$Double\, traverse\, /\, minute = R = \frac{N'}{S} = n\frac{A}{B}\cdot\frac{1}{S}$$

$$So, \tan\theta = \frac{2LR}{\pi dn} = \frac{2L}{\pi d}\times\frac{n\times\dfrac{A}{B}\times\dfrac{1}{S}}{n}$$

$$= \frac{2L}{\pi d}\times\frac{A}{B}\times\frac{1}{S}\propto\frac{1}{d}$$

As, d increases with the package building, the angle of wind decreases. It is also understood from the above expression that $d\tan\theta$ remains constant for spindle-driven winders.

$$Traverse\, ratio = wind\, /\, double\, traverse$$

$$= \frac{wind\, /\, \min}{double\, traverse\, /\, \min} = \frac{n}{R} = \frac{n}{n.\dfrac{A}{B}\cdot\dfrac{1}{S}} = \frac{B\times S}{A} = constant$$

So, for spindle-driven winders, traverse ratio remains constant during the package building.

$$Winding\, speed = \sqrt{(\pi dn)^2 + (2LR)^2}\ (generally\, increases\, with\ `d')$$

Figure below depicts the two situations with low and high package diameters. The traverse ration is same in both the cases. However, the angle of wind has reduced from θ to α.

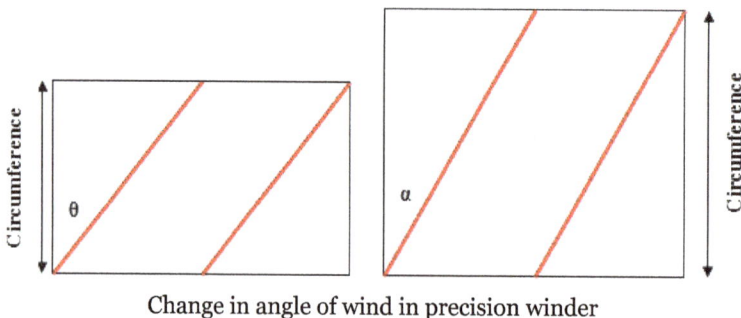

Change in angle of wind in precision winder

Step Precision Winder or Digicone Winder

In step precision winder the problem of patterning is prevented by changing the traverse speed proportionately with the package speed (r.p.m.) so that the traverse ratio value remains constant over a period of time. As the package diameter increases, the package r.p.m. decreases as the package is driven by the drum. The traverse speed is also reduced in the same rate. However, after certain time the traverse speed is raised back to the original value in one step thereby moving quickly from one convenient value of traverse ratio to another. The schematic representation of the drive system of step precision winder is shown in figure.

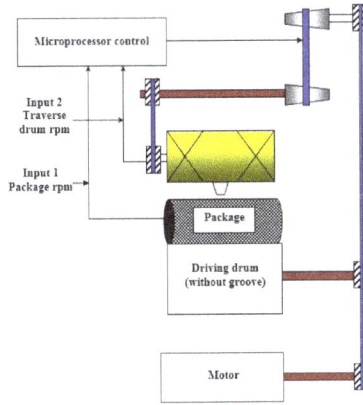

Step precision winder

The driving drum gets motion directly from the motor. However, the motion goes to the traverse guide through a cone drum combinations. As the package r.p.m. reduces, the belt connecting the two cone drums are also shifted towards the left side in a controlled manner. Therefore, the actuating diameter of the driver cone drum reduces whereas it increases for the driven cone drum. So, the speed of the traversing system decreases and traverse ratio remains constant over a period of time. However, the traverse speed cannot be reduced continuously as it will reduce the winding speed and angle of wind continuously. Therefore, after a certain time the connecting belt of the cone drums is shifted towards the right to restore the original value of traverse speed. Therefore, the traverse ratio, governed by the following equation, reduces in steps from one convenient value to the other.

$$Winding\ per\ double\ traverse = \frac{r.p.\text{m of package}}{Double\ traverse\,/\min}$$

Figure depicts the change in traverse ratio with the increase in package diameter.

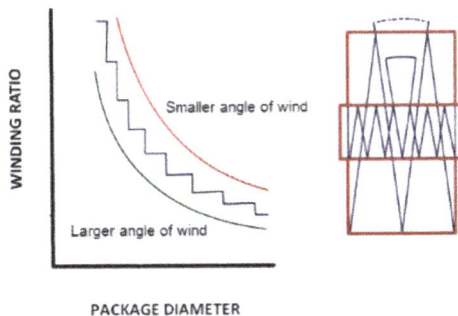

PACKAGE DIAMETER

Change in traverse ratio with package diameter in Digicone winder

Angle of wind (θ) also changes by a small amount (1-2°) when V_t is gradually reduced but again it regain the original value when the V_t is raised back to the original value.

Table: Comparison matrix of winding principles

Parameter	Drum-driven	Spindle-driven	Digicone winder
Angle of wind	Remains constant	Decreases with increase in package diameter	Varies within a very small rang
Traverse ration	Decreases with increase in package diameter	Remains constant	Remains constants for some time and then decreases in step
Winding speed	Remains constant	Generally increases with package building	Reduces slowly due to the reduction in traverse speed and then increases to the original value
Package density	Increases drastically at the zone of ribbons	Increases with the increase in package diameter	Density does not change with package diameter

Gain

Gain is the distance by which the winding point has to be shifted for avoiding patterning. Linear gain is measured in the direction of perpendicular to the direction of package axis as shown in figure. Traverse ratio basically quantifies the number of package revolution within a certain time (one double traverse). Therefore, linear gain cannot be added or subtracted with the traverse ratio. However, linear gain can be divided with the package circumference to obtain revolution gain which can be added or subtracted with traverse ratio.

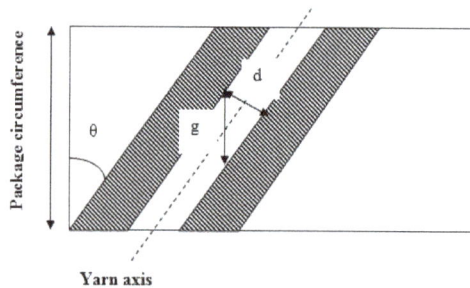

Gain for cheese package

If winding takes place in such a manner that the coils wound in two consecutive double traverse is touching each other physically then the gain can be expressed by the following equation.

$$Linear\ gain = g = \frac{Yam\ diameter}{\sin \theta}$$

where θ is angle of wind.

Pirn Winders

Pirns are the yarn packages used within the shuttle to supply the yarns for pick insertion during weaving. The dimension of the shuttle is restricted by the shed geometry and the strain imposed on the warp yarns during shedding operation. The dimension of the pirn is governed by the dimension of the shuttle. Thus the pirn has to be a long and thin package. In contrast to cone winding, where the supply packages (ringframe bobbins) are small and the delivery packages are big, the supply packages are bigger than the delivery package (pirn) in pirn winding. As the yarns are already been cleaned from slubs and other objectionable faults, no yarn clearing operation is required in pirn winding.

The winding principle of pirn is different than that of cones and cheeses. If a cross-wound package is made then there will be lot of tension variation during weaving. On the other hand, the parallel would package will give the problem of instability. Therefore, pirns are made by overlapping short, conical and cross-wound sections as shown in the figure.

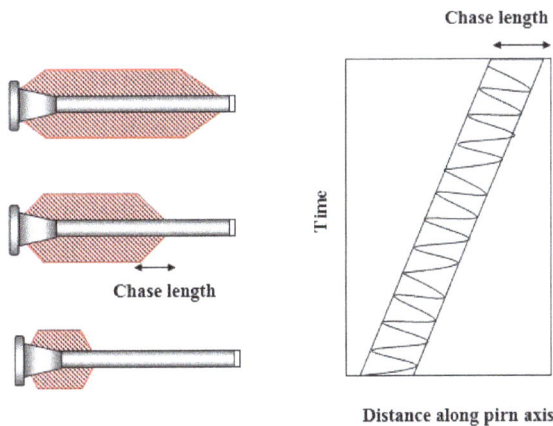

Stages of pirn winding

The base of the empty pirn is generally conical. The pirn winding starts from the conical base and progressively proceeds towards the tip of the pirn. The distance travelled in one stroke of traverse is known as chase length. One layer of coils are laid on the conical base during the forward and as well as duirng the return movement of the traverse meachanism. Thus the conicity of the package is maintained and thus the tip of the cone formed by the coils of yarn slowly proceeds towrds the tip of the pirn. The process can be visualised as if one plastic cup (having cone sahpe) is placed over another and the process is continued to build a tall cylindrical column. This is depicted in the figure. For the ease of visibility, large gaps has been maintained between two cones of coils and thus it seems that the overlapping between two layers of coils is very low which is actually not true.

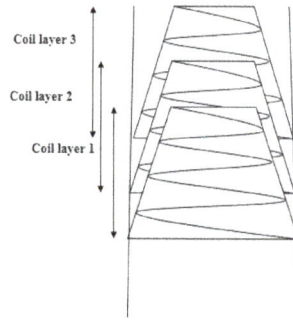

Building of a pirn

Pirns may generally be described in the following categories or types.

- Plain tapered pirn
- Pirn with partly formed (half) base
- Pirn with full base

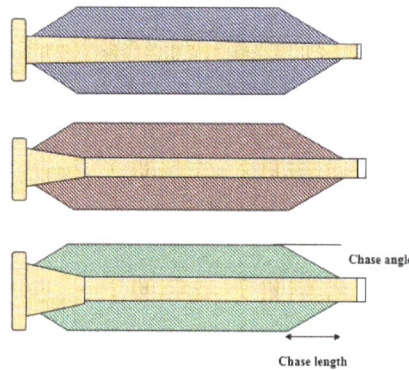

Plain, half base and full base pirns

If the full and empty pirn diameter is D and d respectively, L is the chase length and α is the chase angle, then the following expression can be written.

$$\tan \alpha = \frac{D-d}{2L}$$

The chase angle depends on the type of yarn. For spun yarns the chase angle could be as high as 15-18°. However, for filament yarns with low friction, it could be as low as 6-10°.

Conditions for Uniform Package (Cheese) Building

Uniform building of package is imperative in winding process. The package should have very uniform density. In the following part, the conditions for uniform building of cheese have been derived based on the following assumption.

Assumption: Length of yarn wound per unit surface area of the package should be constant for uniform building of package.

The path of yarn on a cylindrical cheese has been shown in the figure.

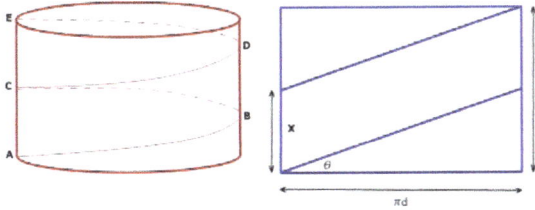

Uniform building of cheese

Diameter of package is d and height of the package is L

Length of one coil $= AC = \dfrac{\pi d}{\cos \theta}$ (θ *is angle of wind*)

Number of such coils in one traverse $= \dfrac{L}{X} = \dfrac{L}{\pi d . \tan \theta}$

Length of yarn / surface area

$$= \dfrac{\text{Total length of yarn wound at diameter } d}{\text{Total surface area of package at diameter } d}$$

$$= \dfrac{\dfrac{\pi d}{\cos \theta} \times \dfrac{L}{\pi d . \tan \theta}}{\pi dL} = \dfrac{1}{\pi d \sin \theta}$$

So, $\pi d \sin \theta$ must be kept const for uniform building of the cheese.

Drum-driven Winder

$$\tan \theta = \dfrac{\text{Traverse speed}}{\text{Surface speed}} = \dfrac{V_t}{\pi n d}$$

where d and n are package diameter and r.p.m.

$V_t \cos \theta = \pi n d Sin\theta$

For uniform building d $\sin \theta$ *should be kept* constant.

For drum driven winder n×d=constant.

$So, n \, \infty \, \dfrac{1}{d}$

$So, V_t \cos \theta \text{ should be changed } \infty \dfrac{1}{d}$

For drum-driven winders with grooved drums, θ remains constant during package building. Therefore, V_t should be reduced with the increase in package diameter d to attain uniform building. This has been shown in the figure.

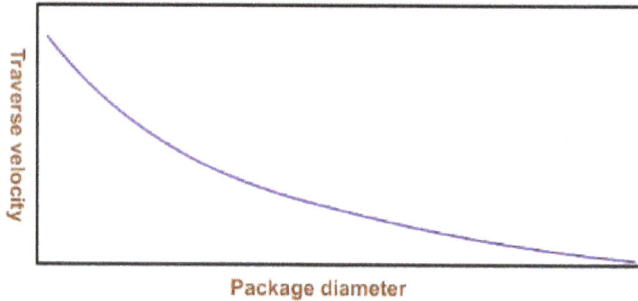

Package diameter vs traverse velocity

Comments

For drum-driven winder, θ is constant provided the ratio of V_t and V_s are constant. But if V_t is reduced (keeping V_s constant theoretically, which is possible if traversing mechanism is separate from groove drum), θ will also reduce and $\cos\theta$ will increase. So, V_t has to be reduced in such a manner that the product of V_t and $\cos\theta$ is changes proportionately with 1/d.

Spindle-driven Winder

For spindle driven machine package r.p.m. i.e is constant So, $V_t \cos\theta$ should be kept constant.

For spindle driven machine θ reduces with the increase of d So, V_t should be reduced accordingly.

Comments

In spindle-driven machine, θ reduces (even when V_t is constant) as package diameter (d) increases. So, V_t has to be reduced accordingly as reducing V_t will have further bearing on θ.

Pirn

A pirn is a rod onto which weft thread is wound for use in weaving. Unlike a bobbin, it is fixed in place, and the thread is delivered off the end of the pirn rather than from the centre. A typical pirn is made of wood or plastic and is slightly tapered for most of its length, flaring out more sharply at the base, which fits over a pin in the shuttle. Pirns are wound from the base forward in order to ensure snag-free delivery of the thread, unlike bobbins, which are wound evenly from end to end.

Wool weaving shuttle, with pirn in middle

Pirns became important with the development of the flying shuttle, though they are also used with other end delivery shuttles. Power looms which use pirns generally have automatic changing mechanisms which removes the spent pirn from the shuttle and replaces it with a fresh one, thus allowing for uninterrupted weaving.

Conditions for Uniform Increase in Cone Diameter

In case of cone, the diameter of package reduces as the yarn traverses from the base to the tip as shown in the figure. Therefore, situation becomes more complicated than the cheese winding. It is important to maintain the conditions so that the diameter in the base and diameter at the tip increases at the same rate. Surface speed of the cone is also less in the tip part as compared to that of base part. The condition for uniform increase of cone diameter has been deduced in the following part.

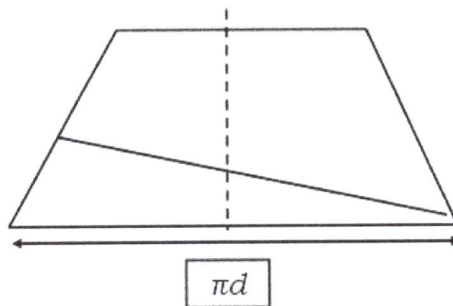

πd

Yarn path on a cone

Two sections of the cones having diameters d_1 and d_2 are being considered. Let w_1, v_1 and s_1 are the winding, traverse and surface speeds respectively at cone section diameter d_1. Similarly, w_2, v_2 and s_2 are the winding, traverse and surface speeds respectively at cone section diameter d2. For the analysis, a small time interval δt is considered.

Length wound per unit surface area at cone diameter $d_1 = \dfrac{w_1 t}{\pi d_1 v_1 \delta t}$

Length wound per unit surface area at cone diameter $d_2 = \dfrac{w_2 t}{\pi d_2 v_2 \delta t}$

For uniform increase in cone diameter, the boundary condition is

$$\frac{w_1 t}{\pi d_1 v_1 \delta t} = \frac{w_2 t}{\pi d_2 v_2 \delta t} \ or \ \frac{w_1}{w_2} = \frac{d_1 v_1}{d_2 v_2}$$

It is known that $\tan\theta = \dfrac{Traverse\,speed}{Surface\,speed} = \dfrac{v}{s}$

Therefore, $\tan\theta_1 = \dfrac{v_1}{s_1}$ *and* $\tan\theta_2 = \dfrac{v_2}{s_2}$

and $w_1^2 = s_1^2 + v_1^2$ *and* $w_2^2 = s_2^2 + v_2^2$

$$So, \frac{w_1^2}{w_2^2} = \frac{s_1^2 + v_1^2}{s_2^2 + v_2^2} = \frac{v_1^2\left(1 + \dfrac{s_1^2}{v_1^2}\right)}{v_2^2\left(1 + \dfrac{s_2^2}{v_2^2}\right)} = \frac{v_1^2\left(1 + \cot^2\theta_1\right)}{v_2^2\left(1 + \cot^2\theta_2\right)}$$

$$= \frac{v_1^2 \, Sin^2\theta_2}{v_2^2 \, Sin^2\theta_1}$$

From boundary condition we know that $\dfrac{w_1}{w_2} = \dfrac{d_1 v_1}{d_2 v_2}$

$$So, \left(\frac{w_1}{w_2}\right)^2 = \left(\frac{d_1 v_1}{d_2 v_2}\right)^2 = \frac{v_1^2 \sin^2\theta_2}{v_2^2 \sin^2\theta_1}$$

or $d_1^2 \sin^2\theta_1 = d_2^2 \sin^2\theta_2$ *or* $d\sin\theta = $ constant

$$\tan\theta = \frac{Traverse\,speed}{Surface\,speed} = \frac{V_t}{\pi dn}$$

where d and n are package diameter and r.p.m respectively

$$V_t \cos\theta = \pi \, dn Sin\theta$$

Here, we are considering only one traverse and therefore package r.p.m. (n) remains constant. For uniform increase of diameter dsinθ, should be constant. Therefore, $V_t$cosθ should be constant during one traverse from base to the tip of the cone.

Comments

As we move towards the tip, the d reduces, so θ increases. As θ increases, cos θ reduces. So we need to increase V_t such that the $V_t$cos θ remains constant. Because increase of V_t will effect on θ also.

Example

A cone is having varying section diameter from base to tip. However, at a particular instance, the entire cone revolves at the same r.p.m. (n). Even if the traverse speed (V_t) is constant, the winding speed changes due to the change in surface speed as the winding point moves from base to the tip. This has been demonstrated pictorially in figure. The surface speed of package reduces as the winding point shifts towards the tip of the cone as $d_1 > d_2$. Therefore, even if the traverse speed remains unchanged, the net winding speed near the base i.e. W_s (d_1) is greater than that of near the tip i.e. W_s (d_2).

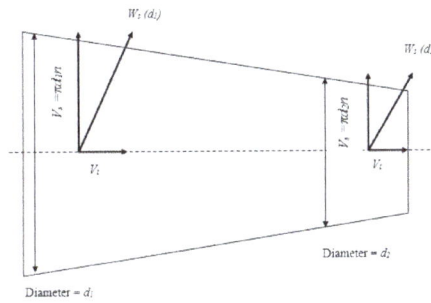

Winding at two different cone section diameters

Let us consider two strips having unit width at cone section diameters d1 (10 unit) and d2 (8 unit) as shown in figure. The length of yarn wound within these strips is a and b respectively (shown by solid portions of inclined lines). So, for uniform increase of cone diameter, the following condition must be maintained.

$$\frac{a}{b} = \frac{d_1}{d_2}$$

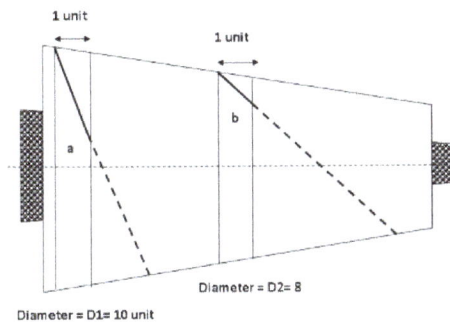

Uniform building of cone

Now, if the length of yarn wound within the strip of cone section diameter d1 is 3 unit, then the length of yarn wound within the strip at cone section diameter d2 can be calculated as follows:

$$b = a.\frac{d_2}{d_1} = 3.\frac{8}{10} = 2.4\,unit$$

As it has been considered that the width of each strip is 1 unit, the angle of wind at different cone section diameter can be calculated as follows.

$$\sin\theta_A = \frac{1}{3}; so\,\theta_A = 19°27'$$

$$V_t\left(at\,diameter\,d_1\right) = V_s\tan\theta_A = \pi.10n.\tan19°27' = 11.09n$$

$$\sin\theta_s = \frac{1}{2.4}; So\,\theta_s = 24°37'$$

$$V_t\left(at\,diameter\,d_2\right) = \pi\,n8.\tan24°37' = 11.52n$$

Now, if we consider that the maximum and minimum cone section diameters are 12 and 6 units respectively, then the angle of wind and traverse speed at different cone section diameters can be calculated as shown in the table. It should be noted that the value of dsinθ is constant irrespective of the cone section diameter.

Table: Calculation of angle of wind and traverse speed at various cone diameters

Cone section diameter (cm)	Angle of wind	Traverse speed (cm/min)	dsinθ(constant)
12	16°8'	10.90n	3.36
10	19°27'	11.09n	3.34
8	24°37'	11.52n	3.35
6	33°45'	12.60n	3.34

Grooves on Winding Drums

The grooves cut on the driving drum of a cone winder have increasing pitch from the base of cone to the nose of the cone so that the traverse speed increases towards the tip of cone.

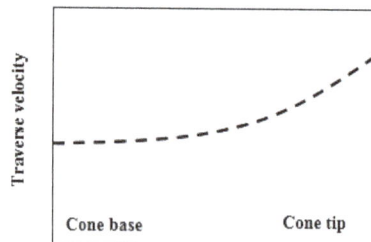

Traverse velocity vs cone section diameter

Yarn Tensioning

The primary objective of yarn tensioning is to build a package with adequate compactness.

Higher yarn tension than the optimum will result in a tighter package and vice versa. If there is any portion of yarn which is very weak from tensile strength point of view (untwisted part of yarn) then it will not be able to sustain the applied winding tension and as a result the yarn will break. This will lead to momentary stoppages in winding operation. However, this will preclude the possibility of yarn breakage in the subsequent processes like warping. As a rule of thumb, yarn tension in winding is around 1 cN/tex .

Types of Tensioning Device

There are two types of tensioning devises used in the winding process.

- Additive type or disc type tensioner

- Multiplicative type tensioner

In case of the former, the yarn is passed through two smooth discs one of which is weighted with the aid of small circular metallic pieces. The weights can be changed easily so that the tension in the output yarn can be adjusted as per the requirement. If T_1 and T_2 are the tension (cN) in the input and output yarns respectively, w is the weight (cN) applied on the top disc and μ is the coefficient of friction between the yarn and metal disc then the following expression can be written.

$$T_2 = T_1 + 2\mu w$$

The factor 2 appears in the above expression as the lower disc also offers reaction forces in the yarn as shown in figure.

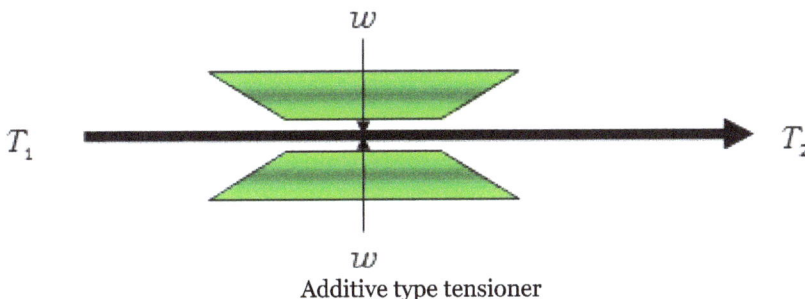

Additive type tensioner

In case of multiplicative type tensioner, the yarn is passed round a curved or cylindrical element as shown in figure.

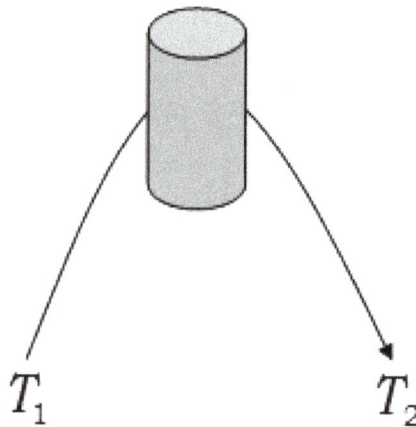

Multiplicative type tensioner

In case of multiplicative tensioner the relationship between input and output tension can be expressed as follows:

$$T_2 \quad T_1 e^{\mu\theta}$$

where θ is the angle of wrap (in radian) of the yarn around the tensioning element.

Figure depicts the two situations with angle of wraps of π and $\pi/2$.

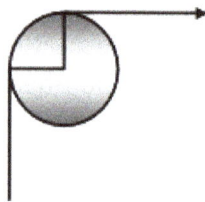

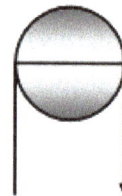

Angle of warp = $\pi/2$ Angle of warp = π

Different angle of warps

Relation between Input and Output Tensions in Multiplicative Tesioner

The yarn is passing over the curvature, shown in red colour, which is considered to be part of a circle (figure). The contact region between the curvature and yarn has created a small angle $d\theta$ at the centre of the assumed circle. The yarn tension in the input side is T_1 and tension in the output side is T_2. The difference between T_2 and T_1 is dT. The difference between the horizontal component of T_2 and T_1 will balance the frictional resistance which will depend of coefficient of friction between the yarn and tensioner (μ) and the resultant vertical component of T_2 and T_1.

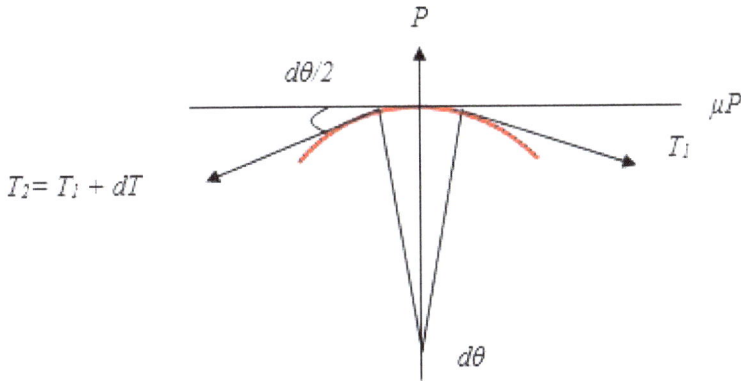

Schematic representation of yarn passing over a curved surface

Balancing the forces, the following equations can be written.

Resolving and balancing the vertical components,

$$P = \left(T_1 + dT_1\right)\sin\frac{d\theta}{2} + T_1\sin\frac{d\theta}{2}$$

$$\cong 2T_1\sin\frac{d\theta}{2}\left(as\;\frac{d\theta}{2}\;is\;small, \sin\frac{d\theta}{2} = \frac{d\theta}{2}\;and\;product\;of\;\frac{d\theta}{2}\;and\;dT_1\;can\;be\;ignored\right)$$

$$\cong 2T_1 \times \frac{d\theta}{2} = T_1 d\theta$$

Resolving and balancing the *horizontal* components,

$$\mu P = \left(T_1 + dT_1\right)\cos\frac{d\theta}{2} - T_1\cos\frac{d\theta}{2}$$

$$= dT_1\cos\frac{d\theta}{2} \cong dT_1\left(as\;\frac{d\theta}{2}\;is\;very\;small\;,\;\cos\frac{d\theta}{2}\cong1\right)$$

$Now\;\mu P = dT_1$

$or\;,\;\mu T_1 d\theta = dT_1$

$$or\;,\;\mu\int_0^\theta d\theta = \int_{T_1}^{T_2}\frac{dT_1}{T_1}$$

$$or\;,\;\mu\theta = \log\frac{T_2}{T_1}$$

$$or\;,\;\frac{T_2}{T_1} = e^{\mu\theta}$$

Numerical Example

1. The tensioning system shown in figure is being used in a winding system. The input and output tensions are 10 cN and 98 cN respectively. If disc (additive) type tensioners A and B are identical then calculate the weights used in tensioners A and B

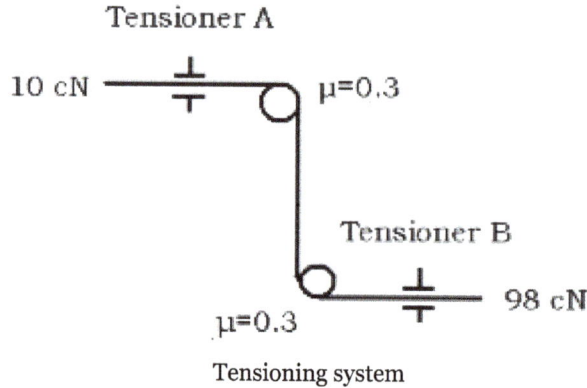

Tensioner A

10 cN μ=0.3

Tensioner B

μ=0.3   98 cN

Tensioning system

Solution

Let the weight of the discs is N cN

$Input\ tension = T_o = 10cN$

$Tension\ in\ yarn\ after\ tensioner\ A = T_1 = T_0 + 2\mu N = 10 + 2\mu N$

$Tension\ after\ the\ first\ multiplicative\ tensioner = T_2 = T_1 e^{\mu\theta}$

$Here, \theta = 90^0 = \dfrac{\pi}{2}$

$Tension\ after\ the\ second\ multiplicative\ tensioner = T_3 = T_2 e^{\mu\theta} = \left(T_1 e^{\mu\theta}\right)e^{\mu\theta}$

$$= T_1 e^{2\mu\theta}$$

$Tension\ after\ tensioner\ B = T_4 = T_3 + 2\mu N$

$$= T_1 e^{2\mu\theta} + 2\mu N$$

$Now = T_1 e^{2\mu\theta} + 2\mu N = 98$

$or, (10 + 2\mu N)e^{2\mu\theta} + 2\mu N = 98$

$or, (10 + 0.6N)e^{0.3\pi} + 0.6N = 98$

$or, 2.56(10 + 0.6N) + 0.6N = 98$

$or, N = 33.9 \approx 34$

So, the weight of each of the discs is 34 cN.

Tension Variation Duirng Unwinding from Cop Build Package

During the unwinding of yarns from cop build packages (ringframe bibbin, pirn etc) short term and long term tension variation is noticed. Short term tension variation arises due to the movement of the yarn from the tip to the base and vice versa (figure a). On the other hand, long term tension variation occurres due to the change in height of the balloon formed between the unwinding point and the yarn guide (figure b).

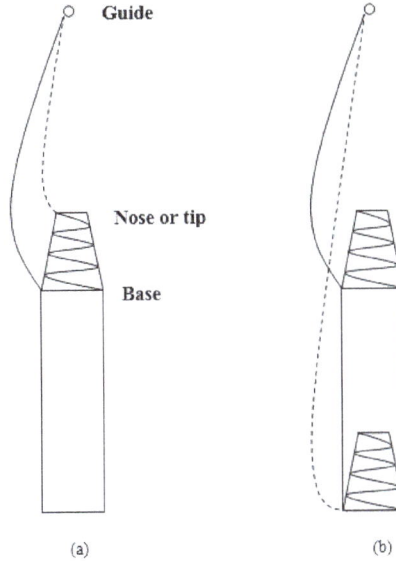

Short term and long term tension variation duirng unwinding

The empirical equation for unwinding tension is given below.

$$Unwinding\ tension\ mv^2 \left[C_1 + C_2 \left(\frac{H}{r} \right)^2 \right]$$

where H is balloon height

r is package radius (varies between tip and base)

m is mass per unit length of yarn

V is unwinding speed

In case of short term tension variation, one layer of coil is unwound and the yarn withdrawal point moves from tip to base, both the H and r increase. However, the proportionate change in r is higher as compared to that of H. Therefore, in every cycle, when the withdrawal point moves from tip to base, the unwinding tension reduces. This has been shown in figure.

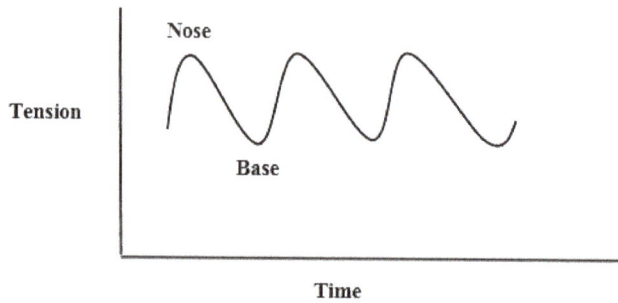

Short tern tension variation

However, over a long period of time, successive conical layers of yarns are removed from the package and thus the conical section of yarns move towards the base of the pirn. Therefore, the balloon height increases resulting in progressive increase in mean unwinding tension. This has been shown in figure.

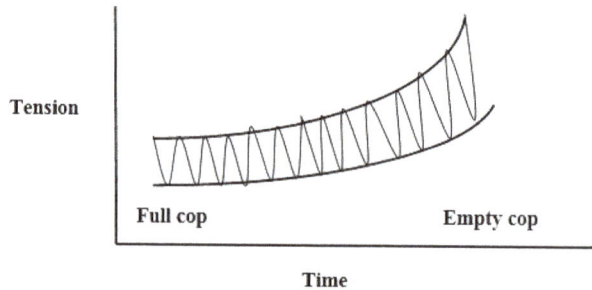

Short term and long term tension variation

Yarn Clearer

A yarn clearer is a device to remove faults (thick places, thin places, foreign matter) from the yarn. Yarn clearing improves the quality of the spun yarn and hence of the cloth made of it. Clear, uniform yarn is especially important for smooth operation of high-speed textile machinery.

Yarn clearing is usually part of the yarn winding step. The yarn from a number of spinning bobbins, called 'cops', is wound on to larger packages called 'cones' for subsequent processing into fabric. During the winding the yarn was traditionally passed through the narrow slit in a steel plate of a yarn clearer or slub-catcher. The object was to catch thick places, or slubs, which occurred when the spinning process suffered an aberration, and to prevent them being woven into the fabric to present unsightly faults.

In modern textile industry, after detecting the faults, the clearer cuts the faulty pieces from the yarn, and after that the piecing device joins the cut ends.

Yarn Clearing

The objective of yarn clearing is to remove objectionable faults from the supply package.

Ideally all the faults present in the yarn should be removed during the yarn clearing operation. However, a compromise is needed and only those faults which have potential to disrupt the subsequent operations or spoil the fabric appearance are attempted for removal during the winding operation. The compromise is done due to the following reasons:

- Removal of yarn faults during winding is associated with the machine stoppages which reduces the machine efficiency.

- When a yarn fault is removed, the yarns are joined again by the knotting or splicing operation which actually introduces a new blemish in the yarn as the strength and appearance of the knotted or spliced region is not at par with the normal region of yarn.

Principles of Measurement

Two principles are used in modern winders for the identification of yarn faults.

- Capacitance principle
- Optical principle

Both the principles have their inherent advantages and limitations. Capacitance system are based on the measurement of yarn mass at a given test length. In contrast the optical systems are based on the diameter measurement.

Figure depicts the principle of a capacitance based yarn clearing system. The yarn is passed at a constant velocity through two parallel plate capacitors. The expression of capacitance of a metallic parallel plate capacitor is as follows.

$$\text{Capacitance} = C = \frac{\varepsilon A}{d} = \frac{k\varepsilon_0 A}{d}$$

where A is the area of the places, d is the distance between the plates, ε is the permittivity of the medium present between the places, ε_0 is the permittivity of vacuum and k is the dielectric constant of the medium.

The permittivity of vacuum (ε_0) is 8.85×10^{-12} F/m.

When the yarn will pass through the parallel plates, the equation will take the following form.

$$C = \frac{A}{\left(\dfrac{d_1}{\varepsilon_1} + \dfrac{d_2}{\varepsilon_2}\right)} = \frac{A\varepsilon_0}{\left(\dfrac{d_1}{k_1} + \dfrac{d_2}{k_2}\right)}$$

where d_1 is the thickness of material 1 (yarn)

d_2 is the thickness of material 2 (air)

ε_1 is the permittivity of material 1

$\varepsilon_2 (\approx \varepsilon_0)$ is the permittivity of material 2

k_1 is the dielectric constant of material 1

k_2 is the dielrctric constant of material 2

Based on the mass of the yarn present within the parallel plate capacitor, the capacitance changes which is converted into mass unevenness. The dielectric constant of water is 80 whereas for textile fibres it is around 2-5 and for air it is nearly 1. Thus the measurement is highly sensitive to the presence of moisture and therefore conditioning of samples in standard atmospheric conditions is of paramount importance.

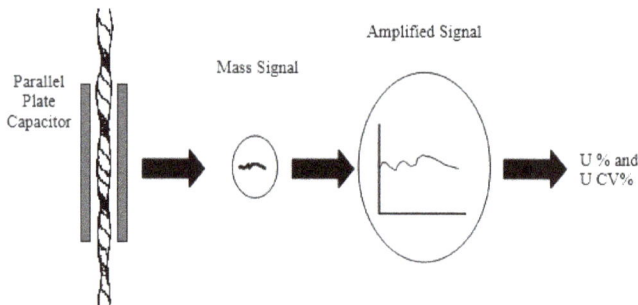

Principle of capacitance based unevenness measurement

Figure below represents the principle of optical based yarn clearing system. The emitter emits light and the receiver detects it and converts to proportional electrical signal. The light received by the receiver will obviously depend on the diameter of the yarn passing between emitter and receiver.

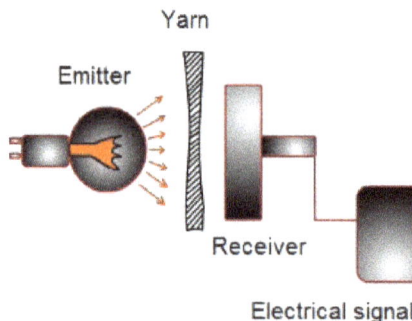

Principle of optical based measurement

It should be considered that 10% deviation in diameter will actually cause 21% deviation in the mass. Because for circular yarn cross-section, $mass \propto diameter^2$. Therefore, principally capacitance based measurements are more sensitive to deviation than the optical based measurements.

However, some of the faults which may not be detected by the capacitance based testers can be detected by optical type testers. The fault may be a low twisted region or a hole within the yarn structure (figure below). If the mass per unit length is same, capacitance type testers will not detect the yarn irregularity although there is a deviation in diameter that can produce fabric defects. This can be detected by optical type testers as higher yarn diameter will reduced the amount of light received by the receiver.

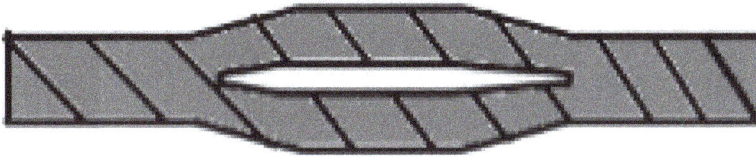

Yarn with a hole within the structure

Yarn blemishes are broadly categorized under two heads

- Frequently occurring type or yarn imperfections

- Seldom occurring type or yarn faults.

Yarn Imperfections

These blemishes occur very frequently in the spun yarns. However, they do not pose serious threat to the subsequent processes or fabric appearance. Frequently occurring faults are measured by yarn unevenness testers and expressed by the frequency of occurrences per km.

1. Thick places (mass exceeds by at least + 50% of the nominal mass)

2. Thin places (mass is lower than by at least - 50% of the nominal mass)

3. Neps (mass exceeds by + 200% of the nominal mass with reference length of 1 mm).

Yarn Faults

Yarn faults are seldom occurring mass variation in the yarn. They can adversely affect the running performance of the loom due to frequent breakage. Besides they can severely damage the appearance of the fabric. Yarn faults are tested by Classimat III or Classimat IV testers and categorized into different classes (23 and 33 respectively) depending on the length and diameter of the faulty place. Yarn faults generally expressed by the number of occurrences per 100 kilometers. Figure shows the matrix of Classimat

III faults where diameter and length of faults are indicated in the vertical and horizontal direction respectively.

The Classimat III faults are classified under three major categories:

- Short thick faults: A1 to D4

- Long thick faults: E, F and G

- Long thin faults: H1, I1, H2, I2

The classifications A, B, C, and D correspond to fault reference lengths of 0.1-1, 1-2, 2-4 and 4-8 cm respectively. The sensitivity % indicates percentage increase in the fault mass varying from +100% to more than +400%, corresponding to diameter increase of 41% and 123%. This results in 16 classifications with A1 the shortest in length and smallest in diameter and D4 the longest in length and largest in diameter. Spinners' double refer to a long thick fault (with the indication E) whose length overstep 8 cm and mass exceeds +100%. F and G are also long thick faults as their mass exceeds the nominal level by + 45% and length is between 8-32 cm and greater than 32 cm respectively. Within the long thin category, H faults are having length between 8-32 cm whereas I faults are longer than 32 cm.

A4, B4, C4, D4, C3 and D3 are generally considered as objectionable faults as their length and deviation from the nominal mass are very high. Now, A3, B3, C2 and D2 are also considered as objectionable faults for high quality products.

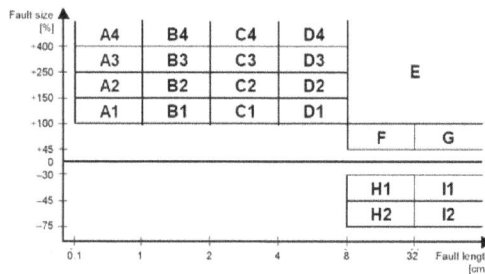

Matrix of Classimat III faults

Figure below shows the matrix of Classimat IV faults. In addition to the 23 faults of Classimat III, there are 10 additional faults. These are as follows:

- Very short thick fault (A0, B0, C0 and D0)

- Short thin faults (TB1, TC1, TD1, TB2, TC2, TD2)

For very short thick faults (A0-D0), the mass exceeds the nominal mass by + 70%. The reference length of these faults is same with A-D which has been mentioned earlier. For TB1, TC1 and TC1, the mass is lower than the nominal mass by -30-45% whereas for TB2, TC2 and TD2 this is -45-75%.

Research work by Aggarwal , Hari, and Subramanian showed that C3, C4 and all D Classimat faults, even after sizing, have lower tenacity, reduced extensibility and poor abrasion resistance. These Classimat faults introduce a very high frequency of low strength and low elongation portions into the yarn, which in turn causes warp breaks.

	A4		B4		C4		D4		E
	0		0		0		0		
A3	0	B3	0	C3	0	D3	0		
A2	30	B2	0	C2	4	D2	0		1
A1	162	B1	8	C1	0	D1	0		
A0	3641	B0	98	C0	15	D0	1	F 0	B 0
T01	23902	T21	5644	T01	337	H1	0	I1	0
T02	485	T02	319	T02	48	H2	0	I2	0

Matrix of Classimat IV faults

Causes of Classimat Faults

The short thick faults (A1-D4) of Classimat system are either caused by the raw material defects and preparatory stages or due to the drafting defects. If a diagonal is drawn joining A4 and D1 then the matrix is divided into two equal parts as shown in figure. According to the thumb rule of spinning, the faults lying within the upper triangle are due to the drafting faults whereas the faults lying below the diagonal are either due to the deficiency in the raw material or due to the opening problem of the blowroom and carding machines.

Fault size [%]				
+400	A4	B4	C4	D4
+250	A3	B3	C3	D3	E
+150	A2	B2	C2	D2
+100 / +70	A1	B1	C1	D1
+45 / 0	A0	B0	C0	D0	F	G
-30				
-45				H1	I1
-75				H2	I2

0.1   1   2   4   8   32   Fault length [cm]

☐ Faults due to raw materials, blowroom & carding

▨ Faults due to drafting systems

Causes of short thick faults

Settings of Yarn Clearing Channels on Winding Machine

In modern winding machines, three channels are used for the clearing the yarn faults.

- Short thick faults: S channel

- Long thick faults: L channel

- Long thin faults: T channel

The user has the flexibility to manipulate the settings of these three channels to optimise the yarn clearing in terms of fault removal and machine stoppages (winding cuts/

km). The setting depends on yarn count as well as quality (carded or combed). Typical settings for SLT channels can be as follows:

- S channel: Mass + 140 to + 200 % and length 1.5-2 cm

- L channel: Mass + 40 to + 50 % and length 40-50 cm

- T channel: Mass -30 to -40 % and length 40-50 cm

Removal of Foreign and Coloured Fibres

The contamination in cotton fibre is one of the most serious causes of quality problems. The contamination includes jute, polypropylene, husk, leaf, hair and paper. Presence of foreign fibres like polypropylene or polyethylene in cotton yarns creates quality problem after dyeing. Modern blworooms are equipped with systems like Vision Shield® which removes the contaminated portions of cottons using sophisticated image processing technology. Some of the spinning mills prefer manual cotton sorting in addition to the automated systems. The electronic yarn clearers in modern winding machines not only remove the objectionable faults but the foreign fibre channel removes the foreign and coloured fibres also.

Some of the systems used to remove foreign fibres are given below:

- Loepfe Zenit Yarn Master®

- Uster Quantum 2

Loepfe Zenit uses Tribo-electric effect, which is a type of contact electrification, for foreign fibre identification. According to this principle, certain materials become electrically charged after they come into contact with another different material and are then separated. Materials have a specific order of the polarity of charge separation when they are touched or abraded with another object. A material which acquires higher position the series, when touched to a material near the bottom of the series, will attain positive charge. The further away two materials are from each other on the series, the greater the charge transferred.

Uster Quantum also presents the distribution of foreign fibres in a matrix based on % deviation in diameter and fault length. In case of fine setting, the faults in the classes B1, B2, C1, D1 and E1 are further divided into subclasses as shown in the following figures.

B11+B12+B13+B14=B1

B21+B22=B2

C11+C12=C1

D11+D12=D1

E11+E12=E1

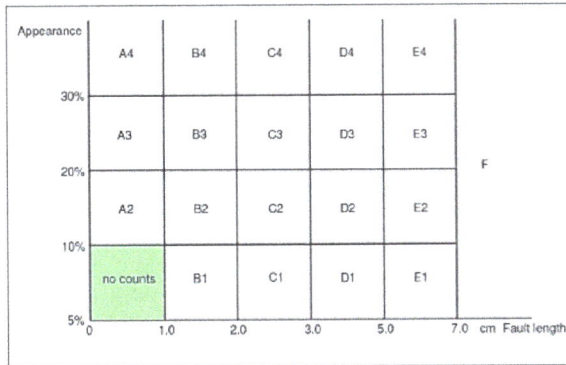

Coarse setting in Uster Quantum 2

No counting is done in the lowest block of left hand side as the number of faults in this category is very high. During winding, only the serious types of foreign fibre faults are attempted to remove. Otherwise the winding efficiency may drop drastically.

Experiments with foreign fibre removal systems have revealed that their clearing efficiency is dependent on the colour of the foreign fibre and the shade depth. The clearing efficiency was found to be higher for red followed by blue and yellow. The clearing efficiency was found to increase with shade depth.

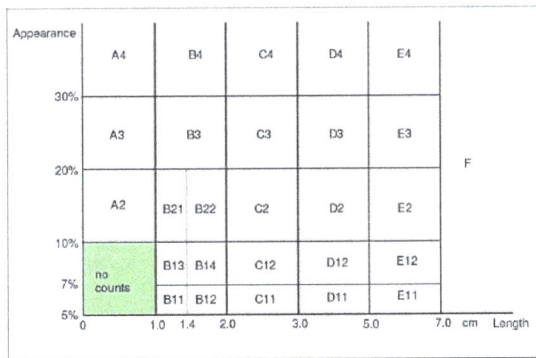

Fine setting in Uster Quantum 2

Splicing

Splicing is the process by which the two ends of yarns are joined. In most of the machines dealing with spun yarns, pneumatic splicers are used. Robotic arms aided with air suction bring two ends of the yarns inside the splicing chamber. Then compressed air is jetted to create turbulence inside the chamber so that the yarn is untwisted. Then some fibres are removed from the yarn ends to create wedge shape. Jetting of compressed air is done again to twist the two superimposed ends of yarns. Splicing introduces a less severe fault in the yarn and the appearance of spliced portion of yarn is checked by making the yarn appearance board. The quality of spliced yarn is checked by using the parameters like retained splice strength and splice breaking ratio as defined below:

$$Retained\,splice\,strength\,(\%) = \frac{Strength\,in\,spliced\,yarn}{Strength\,in\,original\,yarn}.100$$

$$Splice\,breaking\,ratio = \frac{No.\,of\,breaks\,in\,splice\,zone\,(\pm 1cm)}{Total\,number\,of\,tests}$$

Higher retained splice strength (85-90%) and lower splice breaking ratio imply good splicing performance.

Steps of yarn splicing

The performance of a splicer is also evaluated by the parameters like clearing efficiency and knot factor. As knotting device is not used in modern winders like Autoconer, the term knot factor may be replaced with splice factor. Higher clearing efficiency and lower splice factor (close to 1) signifies desirable performance of a splicer.

$$Clearing\,efficiency = \frac{Number\,of\,objectionable\,faults\,removed}{Total\,number\,of\,objectionable\,faults\,in\,yarn} \times 100$$

$$Knot\,factor = \frac{Total\,number\,of\,yarn\,clearer\,related\,breaks}{Number\,of\,objectionable\,faults\,removed}$$

Yarn Winding for Package Dyeing

Yarns are often dyed in package form. Yarn packages intended for dyeing should have certain characteristics to facilitate uniform dyeing within and between the packages. The density of the package should be low but uniform. For cotton yarn, the density of package should be around 0.35-0.40 g/c.c. for optimum dyeing performance. Low density will ensure better penetration and flow of dye liquor across the yarn layers. On the other hand, uniform density will ensure that no preferential channel of fluid flow will be developed. Some of the requirements of packages are as follows:

- Density variation within and between the packages should be less than ± 2.5%

- Dye package outer diameter variation should be ± 1mm.

Drum-driven random winders are not preferred for packages intended for dyeing due to patterning problem which may increase the package density drastically. In case of precision winder, the angle of wind reduces as the package diameter increases. Thus the package density increases towards the outer side of the package. In case of step precision winder or digicone winder, the angle of wind varies marginally during winding (± 1°). Thus the density of the package remains nearly constant. Digicone wider produces the best packages for dyeing.

On principle, the density of the package can be varied by following ways:

- Changing the angle of wind

- Changing the distance between neighboring yarns within a layer

- Changing the pressure between the package and drum

- Changing the Winding Tension

High angle of wind produces lower package density (figure a). Lower the distance between the neighbouring yarns within the same layer, higher will be the package density (figure b). In figures a and b, the path of yarn from left to right has been shown in red colour and right to left in blue colour. However, these two parameters cannot be changed independently in drum and spindle-driven winders. For example, in a drum-driven winder, if the angle of wind is changed by using a winding drum having different scroll (S) value, the distance between the two consecutive coils within the same layer will also change. This can be understood from figure a. Besides, in a drum-driven winder, the distance between yarns within a layer reduces with the increase in package diameter whereas it remains constant in case of spindle-driven or precision winder (Figure c).

Higher pressure between package and winding drum and higher winding tension increases the package density. In some of the modern winders, the pressure between the package and winding drum is reduced as the package diameter increases so that the outer layers do not become denser than the inner layers.

Most of the modern winders are having automatic tension control system which minimizes the tension variation which arises due to the variation in the diameter or due to change in the unwinding point on the supply packages. For a given winding speed, this system ensures controlled increase or decrease of winding tension based on the input tension.

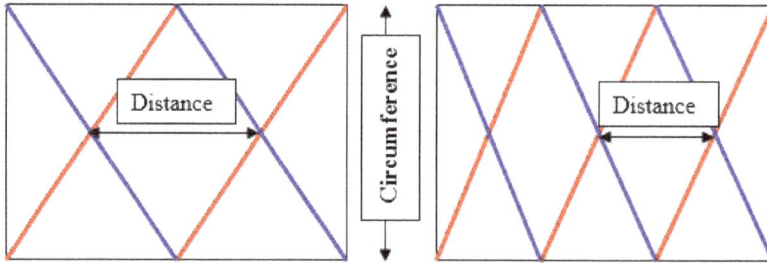

High angle of wind

Low angle of wind

(a) Angle of wind and package density

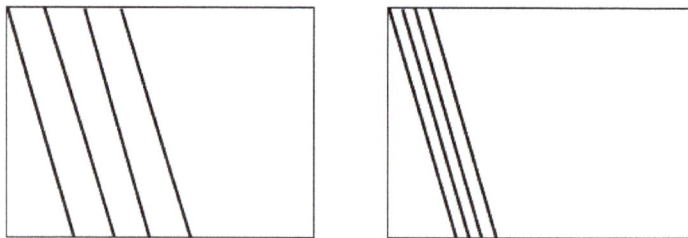

Large gap between yarns

Small gap between yarns

(b) Gap between yarns and package density

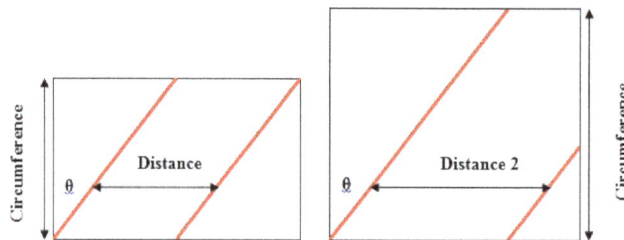

(a) Change in distance between yarns in drum-driven winder

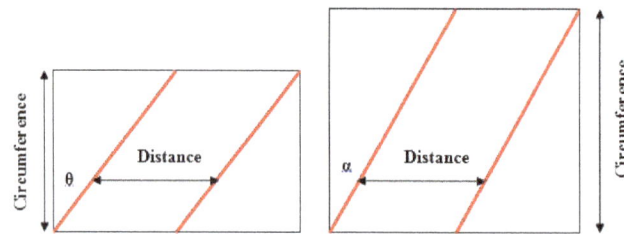

(a) No change in distance between yarns in spindle-driven winder

(c) Gap between yarns at different package diameters

Defects in Winding

Some of the major defects observed in the wound packages are as follows:

•   Ribbon or pattern

- Stitches or jali

- Soft tip or base

- Slough off

The ribbons are formed when the coils of two successive layers rest over or very closely to one another as shown in figure. This happens in drum-driven winder when the traverse ratio becomes an integer. Thick ridges are formed due to patterning and it mars the appearance of the package. Besides, the density of the package at the ribbon part becomes very high which causes problem in unwinding and dyeing.

Patterning is prevented in some of the drum-driven machines by momentarily lifting the package from the drum and thereby creating some intended slippage. This causes change in yarn path from the anticipated one and thus patterning problem is avoided.

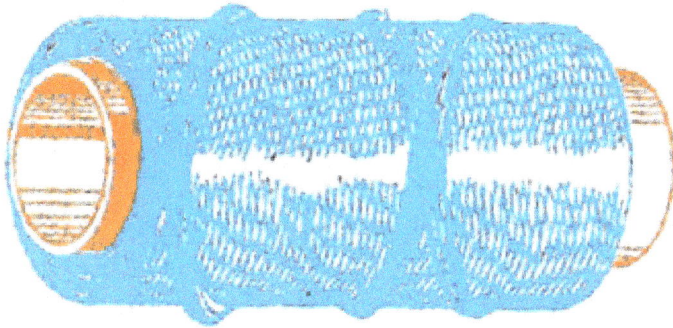

Package with ribbon or pattern defect

Stitches are formed when the yarn is wound beyond the boundary of the package. It may happen due to improper traverse guidance of the yarn at the edges of the package. If the winding tension is low then the yarn may exceed the boundary line up to which it is supposed to travel. The improper contact between the package and the winding drum may also cause stitches.

If the pressure between the package and the drum along the line of contact is not uniform then the package may have different density at the tip and base regions. Higher contact pressure will lead to higher package density and vice versa. The alignment of the winding drum and the package must be correct to ensure uniform density along the axis of the package. The position of the yarn guide should be exactly at the centre of the winding drum so that the length of the yarn between package and the yarn guide remains the same at the two extreme positions of traverse.

Slough off is the problem of removal of multiple coil from the package during high speed unwinding. If the package density or gain is not adequate, then slough off may occur when the wound package will be used in the warping section.

Winding and Yarn Hairiness

The yarn hairiness increases due to winding. The reasons are as follows:

- Abrasion of yarn with various machine parts

- Transfer of one fibres from one section of yarn to the other.

Research by Rust and Peykamian has shown that yarn hairiness increases by a greater amount if the winding speed is high. During winding, redistribution of twist takes place. The twist flows to the regions with lower yarn diameters. This causes fibre transfer and increase in yarn hairiness. Though the average helix angle remains unaltered after winding, the variation in twist angle reduces after winding which supports the fact that twist rearrangement takes place during winding.

References

- Veblen, Sarah (2012). The Complete Photo Guide to Perfect Fitting. Creative Publishing International. pp. 11–2. ISBN 9781589236080

- Camp, Carole Ann (2011). "3: Sewing from a pattern". Teach Yourself VISUALLY Fashion Sewing. John Wiley & Sons. pp. n.p. ISBN 9781118167120

- McGovern, Alyce (6 March 2014). "Knit one, purl one: the mysteries of yarn bombing unravelled". The Conversation. Retrieved 6 March 2014

- Polman, Jan-Willem (2016). Splicing modern ropes - a practical handbook (first ed.). Bloomsbury Publishing Plc. p. 87. ISBN 978-1-4729-2320-2

- Hoffman, Peter J.; Hopewell, Eric S.; Janes, Brian; Sharp, Kent M., Jr. (2011). Precision Machining Technology. Cengage Learning. p. 356. ISBN 1435447670. Retrieved 2013-02-05

- Smee, Sebastian (2009-12-25). "Dave Cole takes knitting to new heights - The Boston Globe". Boston.com. Retrieved 2010-09-01

Warping and Sizing in Textile Manufacturing

The warping process converts the yarn into a usable product which has the requisite number of ends, width and length. Warping is usually done on a frame or loom. Size coating of yarn prevents wear and tear by applying a uniform amount of yarn surface on a loom or beam. The topics discussed in the chapter are of great importance to broaden the existing knowledge on fabric manufacture.

Warping

The objective of warping process is to convert the yarn packages into a warper's beam having desired width and containing requisite number of ends. Uniform tension is maintained on individual yarns during warping.

The yarns are would on the warper's beam in the form of a sheet composed of parallel bands of yarns each coming out from a package placed on the creel. A simplified view of the warping process is shown in the figure below.

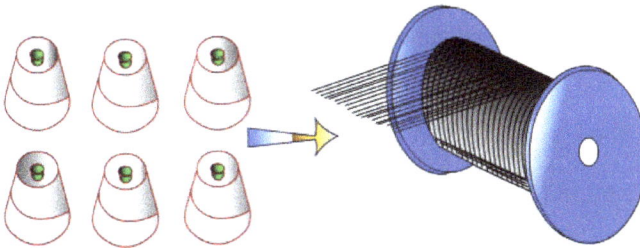

Simplified representation of warping process

Let us take the following hypothetical example to understand the various options of warping process.

A woven fabric roll of 2 m width and 10,000 m length is to be produced from warp yarn of 15 tex. There should be 40 yarns per cm in the fabric. The total number of ends in the fabric will be 40×200 = 8,000. Considering that 10 beams will be combined in sizing, the number of ends on each warper's beam will be 800. Ignoring the yarn crimp and wastage of yarns, the mass of a single end having 10,000 m length will be 150 g. Therefore, there will be the following options:

P × R indicates the number of packages used × number of warping runs

- 800 × 1: Each cone should contain 10,000 m yarn, weighing 150 g.

- 400 × 2: Each cone should contain 20,000 m yarn, weighing 300 g

- 200 × 4: Each cone should contain 40,000 m yarn, weighing 600 g

- 100 × 8: Each cone should contain 80,000 m yarn, weighing 1200 g

- 50 × 16: Each cone should contain 160,000 m yarn, weighing 2400 g

Now, option 1 and option 2 represent two extreme situations. Option 1 (direct warping or beam warping) can be executed when the lot being processed is having significantly higher (15-20 times) length than that of the fabric considered (10,000 m) in this example. Considering the mass of a full cone as 2.1 kg, if the ordered length is 140,000 m, then the entire cone (150 × 14 = 2100 g) will be consumed.

On the other hand, option 2 (sectional warping or indirect warping) is practiced when fancy warp patterns or specialized yarns are used for manufacturing customized products. In this case, the production planning officer does not see the possibility of repeat order in near future. Therefore, he or she wants to consume the entire package to minimize the wastage and inventory carrying cost. Therefore, the beam is made by section by section and the operation is repeated large number of times to complete the entire width of the warper's beam. This is also followed by the beaming operation when all the sections of warp are transferred to a flanged beam.

In synthetic filament weaving systems, each supply package contains a very high length of yarn. Such packages are therefore fairly expensive. It is also very difficult to store such packages with unspent yarn. Hence synthetic yarn weaving units prefer to opt for sectional warping.

For direct warping the typical length of lot can be as follows:

- 40s count: 1.65 lakhs m

- 50s count: 1.80 lakhs m

- 60s count: 1.9 lakhs m

Components of Warping Machine

- Creel (figure)

- Headstock

- Control devices

A simple creel of warping machine

Types of Creel

- Single end creel
- Magazine creel
- Traveling or multiple package creel

Single End Creel

In single end creel, one position of the creel is used for one end on the warper's beam. Single end creel can be of two types namely truck creels and duplicated creels. The creel is movable in case of the former whereas the headstock is movable in case of the latter. In case of truck creel, when the packages from the running creel are exhausted, it is moved sideways and the reserve creel moves into the vacant space (figure). Thus, the time for removing huge number of exhausted package and replenishing them with full packages is saved. However, extra space is required for the reserve creel.

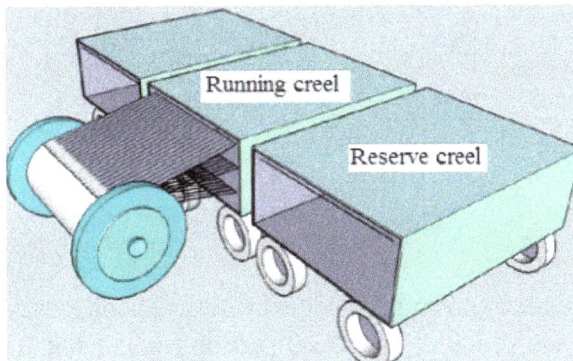

Single end truck creel

Magazine Creel

In magazine creel, the tail end of yarn from one cone is tied with the tip of the yarn of another neighbouring cone. When the first cone is exhausted, the transfer of yarn withdrawal to the second cone takes place automatically and machine does not stop. This has been depicted in the figure. Thus the creeling time is completely eliminated which helps to improve the running efficiency of warping process. However, due to sudden change in unwinding position and tension variation associated with this, some of the yarns break during the transfer (known as transfer failure). The magazine creel has reduced capacity. If the creel has 1000 package holders, then the warp sheet can actually have 500 ends.

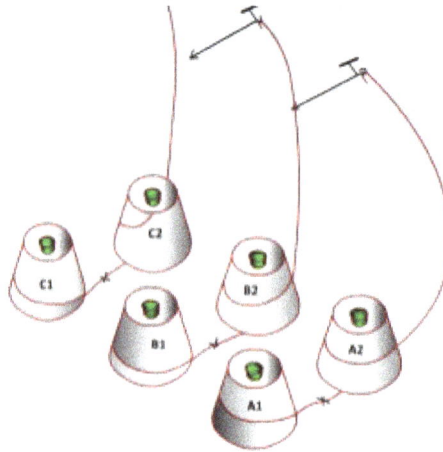

Magazine creel

Travelling or Swivelling Creel

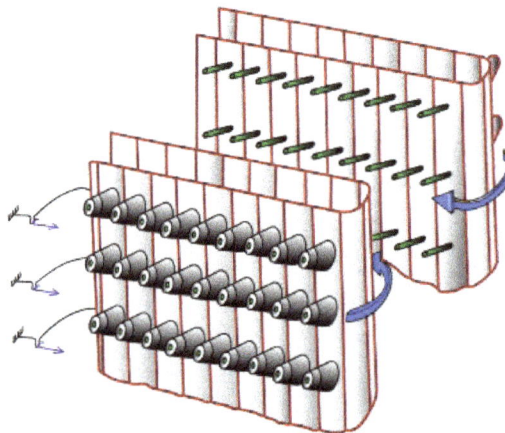

Swivelling creel

In swivelling creel, the pegs (package holders) with full packages can move from inside (reserve) position to the outside (working) position when the running packages are exhausted. Thus considerable time is saved. Then the operator replaces the exhausted packages with full packages when the machine is running. Figure shows the swivelling creel.

Calculation for Warping Efficiency with Different Creels

The yarn mass on full beam is 300 kg, number of ends is 500, yarn count is 30 tex, warping speed is 1000 m/min, cone weight is 2 kg, end break rate 0.1/100 end/ 1000 m, time to repair a break is 0.5 min, Beam doffing time is 5 min, Creeling time is 45 min/ creel, Headstock change time is 3 min/ beam, transfer failure is 2%.

The calculation which follows has been done considering the time required for various operation with respect to one warper's beam. Table shows the time required for various operations.

The mass contributed by a single yarn = 300/500= 0.6 kg

$$The\,length\,of\,the\,warp\,sheet = \frac{0.6\times1000}{30} = 20km = 20,000m$$

Running time = 20,000/1000= 20 minutes.

Creeling time = 45 minutes/ creel.

In case of single end creel, this 45 minutes will be divided between three warper's beam as from one cone of 2 kg mass at least three beams will be made.

So, creeling time assigned to one warper's beam = 15 minutes.

For duplicated creel, the headstock is moved in front of the new creel which is ready with full packages. So, no creeling time is considered. However, it needs the headstock moving time i.e. 3 minutes.

$$Number\,of\,breaks\,in\,warping = \frac{500\times20000\times0.1}{100\times1000} = 10$$

Repair time = Number of breaks × repair time per break =10 × 0.5 = 5 min

Transfer failure = 2 % of 500 ends = 10

Time for repairing the transfer failure = 10× 0.5 = 5 minutes.

This 5 minutes should be equally allocated among multiple warper's beam as from one cone of 2 kg mass at least three beams will be made. So, when the yarns of two cones are tied, six warper's beam can be made without any further creeling.

So, repair time for transfer failure assigned to one warper's beam = 5/6 minutes= 0.83 minutes.

Table: Time required for various operations

Item	Calculation/beam	Single –end (min)	Duplicated(min)	Magazine(min)
Running time	20000 yard sheet	20	20	20
Repair time	0.5 min/break	5	5	5
Beam doffing	5 min/beam	5	5	5
Creeling time	45 min/creel	15	0	0
Headstock change	3 min/headstock	0	3	0
Transfer failure	2%			0.83
Total time		45	33	30.83
Efficiency %		44.44	60.60	64.87

From this example, it can be seen that the type of creel can greatly influence the efficiency of the warping operation.

Drive to Warper's Beam

The warper's beam can be driven by two ways.

- Direct drive

- Indirect drive

In case of direct drive, the warper's beam is driven by gears. As the diameter of the beam increases, the rotational speed of the beam is reduced in order to keep constant warping speed. In case of indirect drive, the warper's beam is rotated by frictional contact with another drum. In this case, the rotational speed of the warper's beam reduces as its diameter increases. Thus the warping speed remains constant.

In modern warping machine (Benninger), the warping speed is around 1200-1400 m/min. The full beam diameter is 1-1.4 m.

Sectional warping is preferred over beam warping for multi-coloured warp. Here the entire width of the warping drum is not developed simultaneously. It is developed section by section as depicted in the figure. As only one section is built at a time, a support is needed at one side of the drum. This is provided by making one side of the drum inclined. The inclination can be of two types.

- Fixed angle

- Variable angle (7 0, 9 0, 11 0 etc.)

As the winding of one layer is completed on the drum, the section of ends is given a requisite traverse so that the end at one extreme corner touches the inclined surface. Thus it gets support from the inclined surface.

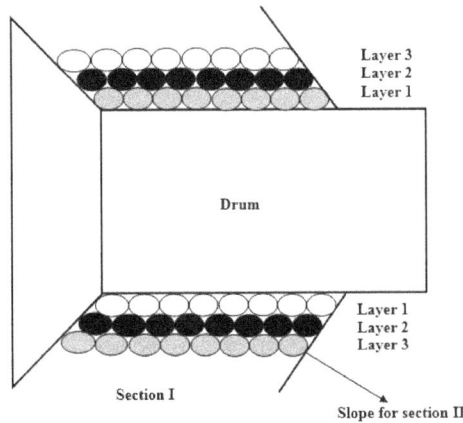

Schematic representation of sectional warping principle

As the process continues, the thickness (or height) of the section gradually increases. When requisite length has been wound in a section, next section is started by shifting the expandable reed assembly by suitable distance (distances of figure).

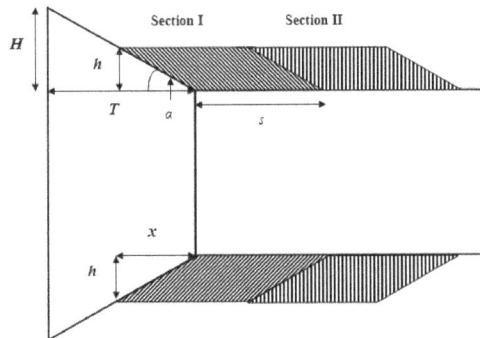

Traverse in sectional warping

If α is the angle of inclination, x is the traverse given to the section and h is the height of the section then

$$\frac{h}{x} = \frac{H}{T} = \tan \alpha$$

$$So, x = \frac{h}{\tan \alpha}$$

For drum with fixed angle, if the yarn is coarser then one layer of the warp ribbon will result in higher increase in thickness (Δh) and thus to match the inclination, the traverse speed (Δx) should be higher. The machines are designed with provisions to change the traverse speed so that a wide range of yarn counts can be managed. For fixed angle drums, only one variable i.e. traverse speed is to be adjusted while with variable angle drums both traverse and the angle of inclination can be varied.

For drums with variable angle, the angle is changed by changing the inclination of metal plates which are supported at the end of the drum. When the angle is increased, larger gaps are created between the neighboring metal plates. Therefore, the yarn will remain unsupported at the gaps between two metal plates.

The flow chart of the warping process can be represented as shown in figure.

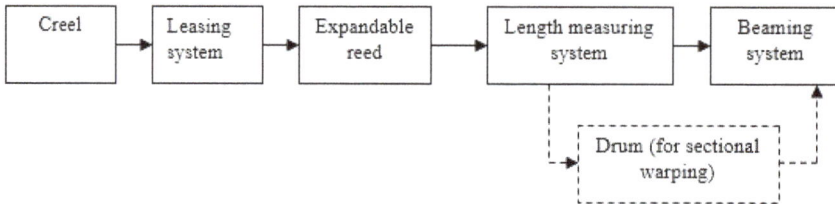

Sequence of warping process

Leasing: It is asystem by which the position of the ends is maintained in the warp sheet. Generally it is done by dividing the ends in two groups (odd and even). If odd ends are passing over the lease rods then the even ends will pass below the rod. The relative positions of the ends will reverse in case of second lease rod.

Expandable reed: It is used to control the spacing between consecutive ends. The two limbs of V shaped expandable reed can be expanded or collapsed as per the required spacing of ends. Figure shows the expandable reed and lease rods in a sectional warping machine.

Beaming system: In the beaming process, all the sections are simultaneously transferred to the flanged warper's beam. The drum is rotated by the tension of warp sheet whereas positive drives are given to the warper's beam. The speed of beaming process in sectional warping is quite slow (around 300 m/min).

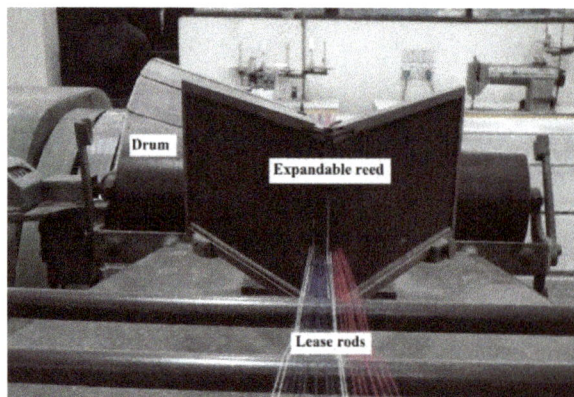

Expandable reed and lease rods

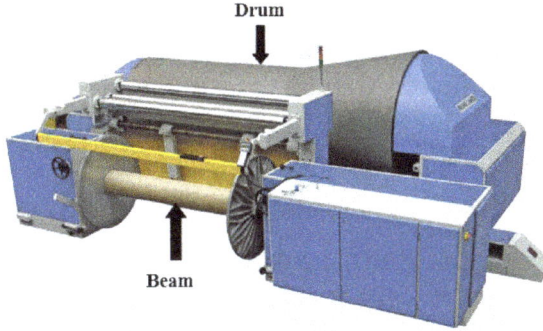

Sectional warping drum and beam

Table: Comparison between beam warping and sectional warping

Beam warping	Sectional warping
Used for high volume production	Used for small volume and customised production (stripes and specialised yarns)
One step process	Two step process
High creel capacity is required	Low creel capacity is sufficient
Comparatively less expensive	Comparatively more expensive
Beaming speed is high	Beaming speed is low
More common	Less common

Warp Stop-motions

Warp stop-motion stops the loom in the event of an end break. The system is activated by the lightweight metallic drop wires which have profiled shape. Two such drop wires or droppers are shown in the figure. The large slot at the top is for the movement of the reciprocating bars which are used in both mechanical and electrical warp stop-motions. Design (a) can be used when a single end is passed through the drop wires during the drawing in operation. Design (b) can be used after the beam gaiting as it has an open ended hole.

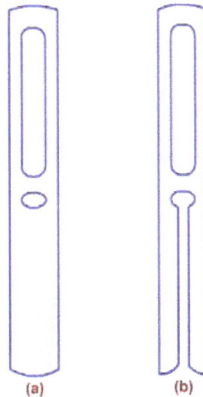

Drop wires

In case of mechanical warp stop-motion, one reciprocating bar moves between two stationary bars. The bars have profiles like step waves. The sideways movement of the centre bar is equal to the width of a step. In case of an end break, the drop wire will lose support from the tight yarn and will fall due to gravity. If it falls to the lowest possible height, then the reciprocating movement of the centre bar will be thwarted and the loom is stopped.

In case of electrical stop-motion, the drop wire acts as an element that makes or breaks an electrical circuit. In case of warp break, the drop wire will complete an electrical circuit and activate a solenoid. The solenoid will attract a bar which will be hit by the knock-off lever. As a result the bar will disengage the starting handle through some other levers. Figure shows the warp stop-motion mounted on a loom.

Warp stop-motion on a loom

Weft Stop-motions

Weft stop-motions stop the loom if the weft carried by the shuttle is broken. It is a very important motion as beat up without a pick will necessitate adjustment of cloth fell position before the restart of the loom. The problem of cloth fell position adjustment will be relatively lower in following cases.

- Coarser yarn count

- Higher yarn hairiness

- Higher pick density

Side Weft Form Motion

Side weft fork motion operates at the left side of the loom at the vicinity of the starting handle of the loom. When the shuttle reaches the shuttle box inserting an unbroken pick, the trail of pick pushes the lower end of the fork as the sley moves forward as depicted in Figure (a). This creates anticlockwise movement in the fork according to Figure (a). However, the movement of the fork will be clockwise according to Figure (b). From Figure (b), it can be understood that the notched hammer moves towards the front of the loom once in two picks as it gets motion from a cam mounted on the bottom shaft. In the absence of weft break, the movement of the fork created by the push exerted by the pick clears the upper end of the fork from the notched hammer when the latter is moving towards the front of the loom. Thus the loom continues to run. In the case of a weft break, the upper end of the fork is caught by the notch of the hammer. So, when the hammer is moving towards the front of the loom, the weft fork support pushes the knock-off lever and the latter dislocates the starting handle to stop the loom. Figure (Top view of side weft fork motion) presents the top view of the side weft fork system mounted on a loom.

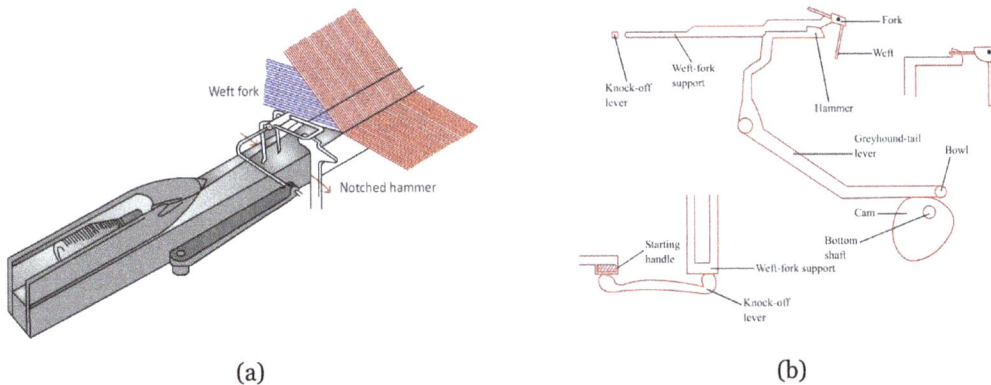

(a)   (b)

(a) Side fork and notched hammer (b) Side view of side weft fork motion

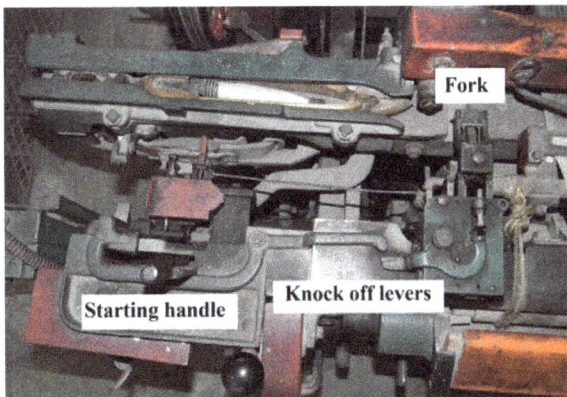

Top view of side weft fork motion

Center Weft Fork Motion

Side weft fork system can detect the weft break after the insertion of one or two missing picks. This problem can be mitigated by using centre weft form motion which is mounted near the middle of the loom. It checks the weft break at every pick and stops the loom before the beat up in case of a weft break. Thus centre weft fork motion is more efficient than the side weft fork motion. Figure shows the centre weft fork.

The centre weft fork is housed in the slot on the sley. The fork rotates clockwise to make a clear passage for the shuttle. This is done by the left sideways movement of the weft fork cam. In the presence of a pick, the fork is supported by the former when the sley moves forward for the beat up. If case of a weft break, the fork looses the support and thus weft fork bowl will be lowered and trapped in a notch restricting the movement of a rod which finally creates the loom stoppage.

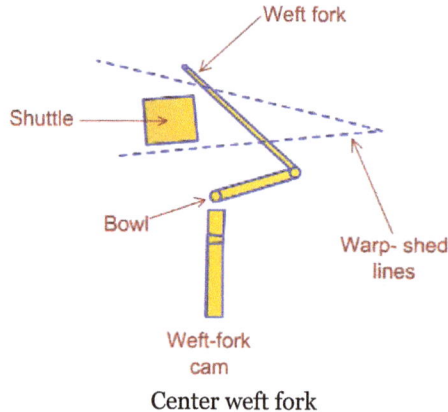

Center weft fork

Warp and Woof

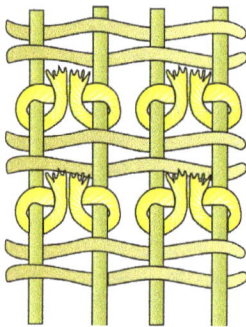

Ghiordes knot

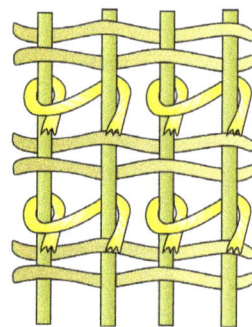

Senneh knot

The yellow yarn is the pile and the horizontal and vertical yarns are the warp and the woof.

In weaving, the weft (sometimes woof) is the thread or yarn which is drawn through, inserted over-and-under, the lengthwise warp yarns that are held in tension on a frame or loom to create cloth. Warp is the lengthwise or longitudinal thread in a roll, while weft is the transverse thread. A single thread of the weft, crossing the warp, is called a *pick*. Terms do vary (for instance, in North America, the weft is sometimes referred to as the *fill* or the *filling yarn*). Each individual warp thread in a fabric is called a *warp end* or *end*.

The weft is a thread or yarn usually made of spun fibre. The original fibres used were wool, flax and cotton. Today, man-made fibres are often used in weaving. Because the weft does not have to be stretched on a loom in the way that the warp is, it can generally be less strong.

The weft is threaded through the warp using a "shuttle", air jets or "rapier grippers". Hand looms were the original weaver's tool, with the shuttle being threaded through alternately raised warps by hand. Inventions during the 18th century spurred the Industrial Revolution, with the "picking stick" and the "flying shuttle" (John Kay, 1733) speeding up production of cloth. The power loom patented by Edmund Cartwright in 1785 allowed sixty picks per minute.

A useful way of remembering which is warp and which is weft is: 'one of them goes from weft to wight'.

Etymology

The words *woof* and *weft* derive ultimately from the Old English word *wefan*, to weave. Warp means "that which is thrown away" (Old English *wearp*, from *weorpan*, to throw, cf. German *werfen*, Dutch *werpen*).

Warp

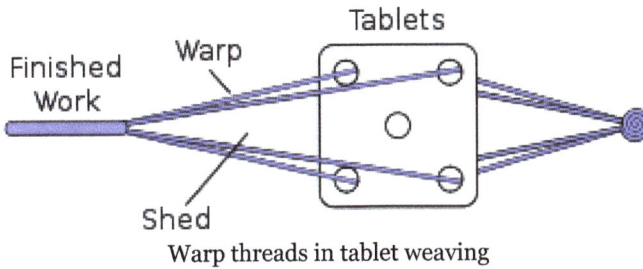

Warp threads in tablet weaving

Very simple looms use a spiral warp, in which a single, very long yarn is wound around a pair of sticks or beams in a spiral pattern to make up the warp.

Because the warp is held under high tension during the entire process of weaving and warp yarn must be strong, yarn for warp ends is usually spun and plied fibre. Traditional fibres for warping are wool, linen, alpaca, and silk. With the improvements in spinning technology during the Industrial Revolution, it became possible to make cotton yarn of sufficient strength to be used as the warp in mechanized weaving. Later, artificial or man-made fibres such as nylon or rayon were employed.

While most people are familiar with weft-faced weavings, it is possible to create warp-faced weavings using densely arranged warp threads. In warp-faced weavings, the design for the textile is in the warp, and so all colors must be decided upon and placed during the first part of the weaving process and cannot be changed. Warp-faced weavings are defined by length-wise stripes and vertical designs due to the limitations of color placement. Many South American cultures, including the ancient Incas and Aymaras used a type of warp-faced weaving called backstrap weaving, which uses the weight of the weaver's body to control the tension of the loom.

Metaphorical Use

The expression "woof and warp" (also "warp and woof", "warp and weft") is sometimes used metaphorically as one might similarly use "fabric"; e.g., "the warp and woof of a student's life" means "the fabric of a student's life." The expression is used as a metaphor for the underlying structure on which something is built. Warp or woof are also words found in the Bible in the discussion of mildews found in cloth materials in Leviticus 13:48-59.

In Hairdressing

Weft is also a hairdressing term for temporary hair extensions. It can be attached to a person's hair in a couple of ways including cornrow braiding, using metal cylinders or gluing. The result is often called a *weave*.

Wrap Reel

A wrap reel on display in Helmshore Mills Textile Museum. This was made by Goodbrand of Manchester to measure cotton.

A wrap reel or skein winder is a device for measuring yarn and making it into hanks of a standard size.. The reel is of a standard size and its revolutions are counted as the yarn is wrapped around it. Typically, a set number of revolutions will be used so that the hank is of a standard size — a skein or lea. For example, a skein of cotton would be 80 turns on a reel of 54 inches circumference, making 120 yards, while the standard length for wool worsted would be 80 yards.

The tension of the yarn as it was wound onto the reel was important because it would be elastic and so a standard tension was required to ensure uniformity. For a given reel, this would be determined by the friction of the setup and so the test hanks would be made and measured in other ways to calibrate the device.

The Science Museum in London has an 18th-century wrap reel in its collection which was made for Richard Arkwright's first cotton mill in Derbyshire. It is kept in their storage archive at Blythe House.

Warp Sizing

The objective of warp sizing is to improve the weaveability of yarns by applying a uniform coating on the yarn surface so that protruding hairs are laid on the yarn surface.

During the weaving process, the warp yarns are subjected to abrasion with various loom components like back rest, heald eyes, reed, front rest etc. In shedding operation, warp yarns also abrade against each other. Size coating protects the yarn structure from abrasion. Therefore, the warp breakage rate in the loom reduces.

Benefits of Sizing

- It prevents the warp yarn breakage due to abrasion with neighbouring yarns or with back rest, heald eye and reed.

- It improves the yarn strength by 10 to 20%, although it is not the primary objective of sizing process.

Characteristics of Sized Yarn

- Higher strength

- Lower elongation

- Higher bending rigidity

- Higher abrasion resistance

- Lower hairiness

- Lower frictional resistance.

Definitions

Size paste concentration, size pick-up and size add-on are some of the terms used frequently in the discussion of sizing process. They are defined as follows:

$$Size\,concentration\,(\%) = \frac{Oven\,dry\,mass\,of\,size\,materials}{Mass\,of\,size\,paste} \times 100 = \frac{S}{P} \times 100$$

$$Size\,add\,on\,(\%) = \frac{Oven\,dry\,mass\,of\,size\,materials}{Oven\,dry\,mass\,of\,unsized\,yarns} \times 100 = \frac{S}{Y} \times 100$$

$$Wet\,pick\,up = \frac{Mass\,of\,size\,paste}{Oven\,dry\,mass\,of\,unsized\,yarns} = \frac{P}{Y}$$

$$= \left(\frac{S}{Y} \times 100\right) \times \left(\frac{P}{S} \times \frac{1}{100}\right)$$

$$= \frac{Add\,on\,(\%)}{concentration\,(\%)}$$

Sizing-weaving Curve

For the sizing process, depending on the size materials used, there is a target add-on for the optimum performance of the warp yarns in the weaving process. This can be understood from the sizing-weaving curve. The solid line represents the warp breakage rate whereas the broken line implies loom efficiency.

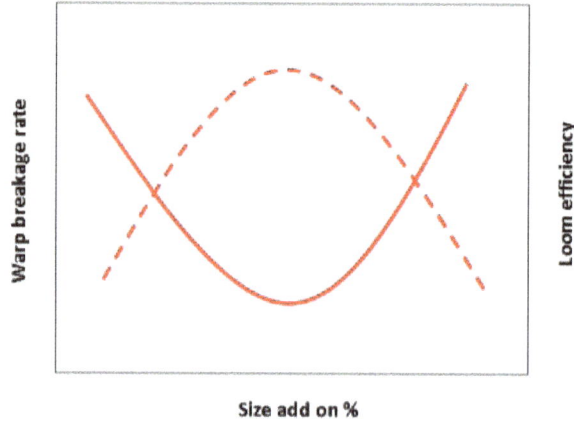

Sizing-weaving curve

At very low level of size add-on, the yarn is not adequately covered by the size film and therefore the yarn is not protected from the abrasion with various loom parts. So, warp breakage rate is generally high at very low level of size add-on. The performance of the yarn in weaving improves as the size add-on increases. The optimum add-on level is marked by very low level of warp breakage rate. However, if the size add-on is higher than the optimum level then warp breakage rate increases again largely due to the loss of elongation and increase in bending rigidity of the yarns.

The optimum level of size add-on will depend on the following factors:

- Type of fibre
- Type of size materials
- Yarn spinning technology
- Yarn count and twist
- Level of hairiness in the yarn
- Loom type and loom speed

Although add-on primarily influences the weaving performance, it is possible to have different weaving performances even at the same level of size add-on. This can happen due to differences in (a) Size penetration and (b) Size coating or encapsulation. This can be explained from the following figure.

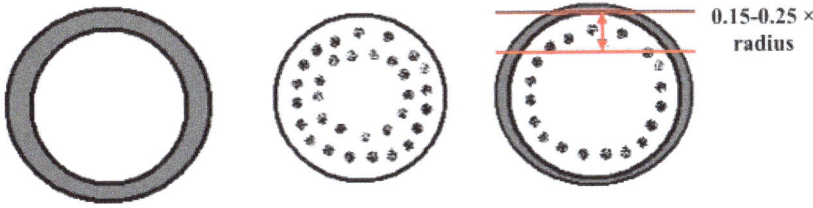

0.15-0.25 × radius

Size coating or encapsulation, size penetration and optimum coating and penetration

In the first case, the size material has formed a uniform coating on the yarn surface. But the penetration of the size material is in adequate. This may tackle the hairiness problem. However, the adhesion of the size film with the yarn will be lower. The size coating will also be very stiff. Therefore, during the abrasion with various machine parts and neighbouring yarns, the size will fall (shed) and thus the weaving performance will be poor.

In the second case, the coating or encapsulation on the yarn surface is inadequate. Therefore, the yarn will not get adequate protection against the abrasion. Besides, the size materials have penetrated too much inside the yarn which is not desirable.

The third case is the optimum one where there is a thin but uniform coating formation on the yarn surface by the size materials. Moreover, the size materials have also penetrated to some extent in the yarn structure which ensures good adhesion between the yarn and the size coating. For optimum weaving performance, size should penetrate to a distance equal to 15-25% of yarn radius.

Desirable Properties of Sizing Materials

The sizing material must fulfil some essential properties and at the same time it is expected that it will have some additional desirable properties. The sizing material must form a smooth and uniform coating the yarn surface. This is known as film forming property. The coating will not only embed the protruding fibres (hairs) on the yarn body but also protect the yarn structure from repeated abrasion during weaving. The size film should adhere with the fibres strongly to prevent shedding (dropping of size film). The film should also have enough flexibility to cope with the flexing or bending of yarns around the back rest, heald eyes and other loom components. The non-exhaustive list of essential and desirable properties is given in the table.

Table: Essential and desirable properties of sizing materials

Film forming	Controllable viscosity
Adhesion	Easy removal and recyclability
Optimum penetration	Neutral pH
Film flexibility and elasticity	Non-polluting
Lubrication	Cheap
Bacterial resistance	

The specific requirements of sizing material properties will depend on the following factors.

- Type of fibre (cotton, viscose, nylon, polyester etc.)

- Type of yarn (ring, rotor, friction, air-jet etc.)

- Type of loom (shuttle, projectile, air-jet, water jet etc)

- Construction features (weave, yarn counts and sett)

Composition of Sizing Material

The specific composition of sizing material depends on the fibre type, yarn type, yarn count, fabric sett etc. However, the materials can be classified under the categories of adhesive, softening agent, antimicrobial agent etc. The adhesive part is responsible for forming the film and adhering with the fibres. Softening agent makes the film flexible so that the film can bend easily without forming cracks. Antimicrobial agents are added to thwart the mildew to grow on the size film. Sometimes, weighting agents and dyes are also added to fulfil specific requirements.

Cotton yarns, in general, are sized by the starch which forms the adhesive component of the size mix. The reason behind the popularity of starch can be attributed to the following factors.

- Starch is chemically same with cotton and rayon and thus the adhesion is very good.

- Desizing is easy

- Relatively cheap

- Properties can be tuned to cope with the need

However, starch gives very stiff film. It has higher biological oxygen demand (BOD). Besides, cooking of starch is required to attain uniformity. Besides, starch has poor bacterial resistance. Overall, the positive attributes of starch dominates over the drawbacks and thus it is still being used in the industry as the primary materials for the sizing of cotton yarns.

The softening materials compensates for the abrasive and harsh feel that is provided by most of the starches. Softeners also lubricate the yarns so that they can pass easily over machines parts without shedding. It also prevents the sticking of size ingredients over the drying cylinders. Mutton tallow which is composed of glycerides of palmitic, stearic and oleic acids is used as softeners. The proportion of softener in the size mix is very crucial as excess use of it deteriorates the strength of size film.

Sources of Starch

Starches are available from the seed, root or pith of plants. Corn, rice and wheat are the examples of seed starch. Potato and Tapioca starches are obtained from roots. Sago starch is obtained from pith. Starches are prepared by grinding the seed, root or pith into fine flour. When the flour is mixed with water and cooked, it produces a thick and smooth glutinous solution. Corn (Maize) starch is the most popular type of starch used in textile sizing. Around 50% of the corn is composed of starch. Corn starch is generally preferred for the sizing of coarse and medium count yarns. Potato yields around 20% starch. It is slow congealing type and therefore gets more chance to penetrate within the yarn structure. It forms a smooth and pliable film on the yarn body. Potato starch is preferred for sizing finer yarns.

Chemical Structure of Starch

Cellulose and Starch

- Chemically same

- Both are polymer of glucose

- Glucose can have two structural (anomeric) form known as α and β.

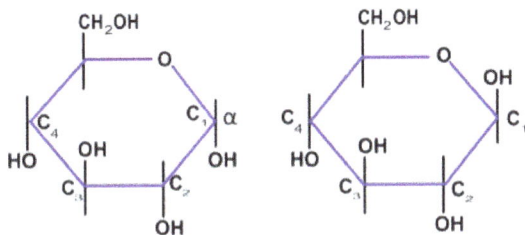

α - glucose   Beta - glucose

Structure of α and β glucose

It can be seen from the above figure that the position of OH group in the pyranose ring at C2, C3 and C4 are same in cellulose and starch. For α -D-glucose, which is the building block of starch, both the OH groups are down with respect to the next glucose molecule. Therefore no change in orientation is needed for condensation as shown in the next figure.

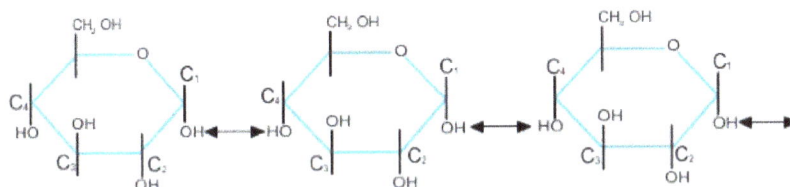

α-glucose   α-glucose   α-glucose

Condensation of α-glucose units

On the other hand, in β glucose, the OH group is up at C1 and down at C4. As a result, every other glucose unit must flip over before the 1,4 hydroxyls can come closer for condensation. This is shown in this figure.

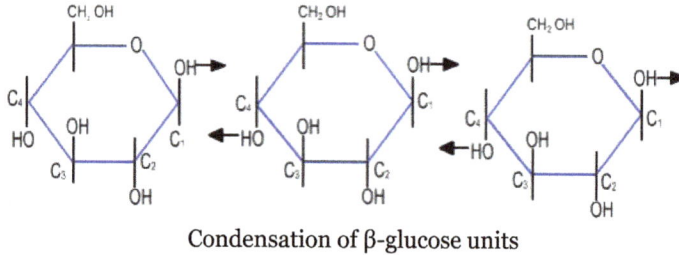

Condensation of β-glucose units

Starches are having two components. The straight chain component is called amylose. The branched chain component is called amylopectin. The differences between amylose and amylopectin have been highlighted in the table.

Table: Differences between amylose and amylopectin

Amylose	Amylopectin
Provides strength	Prevents rapid gelling
Water soluble	Water insoluble
Low molecular weight	Relatively high molecular weight
20-30%	70-80%

The proportion of amylose and amylopection differs depending on source. For example, in potato starch the ratio is 20:80 whereas in wheat starch the ratio is 25: 75.

Cooking of Starch

Starch remains tightly bound in granules and therefore it does not act as adhesive in cold water. Cooking of starch is required to make it soluble in water. The change in viscosity of starch solution during cooking is shown in the figure.

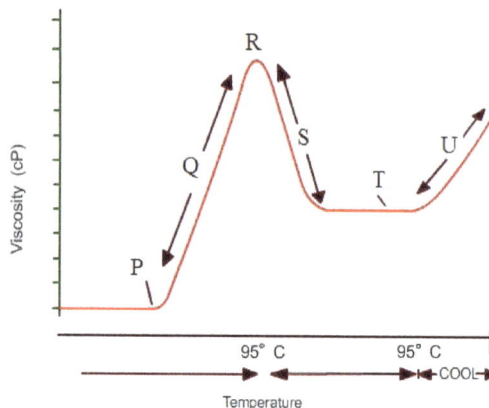

Cooking of starch

Within the granule, the chain molecules of amylose and amylopection are arranged radially in stratified layers. External heat energy is required for the penetration of water molecule within the structure. The temperature at which the thermal energy becomes sufficient to overcome hydrogen bonding within the structure is called 'gelatinisation' temperature (P). Crystallization of starch is lost during gelation. As the water penetrates, the chain molecules are pushed away from each other causing swelling of the starch granule. This is marked by increased in viscosity of the solution (Q). This continues up to the point R. Aided by the continuous shearing provided by the stirring, the starch granules finally break. The chain molecules of amylose and amylopection come out within the solution causing reduction in viscosity (T). When all the granules have burst, the viscosity stabilizes or levels off (T). When the solution is cooled, the starch gels due to the formation of a rigid interlocked micelle-like structure having hydrogen bonding (U). This gel form of starch can form a continuous coating on the yarn surface.

Acid Treatment of Starch

The viscosity of the sizing paste influences the wet pick-up and resultant add-on %. The viscosity is influenced by the concentration of starch (solid content) and molecular chain length of starch. To reduce the concentration of the sizing paste, keeping the solid content same, acid treatment is performed. Aqueous solution of starch is treated with hydrochloric acids at specified temperature and duration. The acid cleaves the polymer at the glycoside linkage and thus the length of the polymer chain is reduced (figure below). Hence, the viscosity is reduced and the fluidity, which is the reciprocal of viscosity, increases. The acid treated starch is often termed as 'thin boiling starch' as it results lower viscosity than the normal starch at a given concentration.

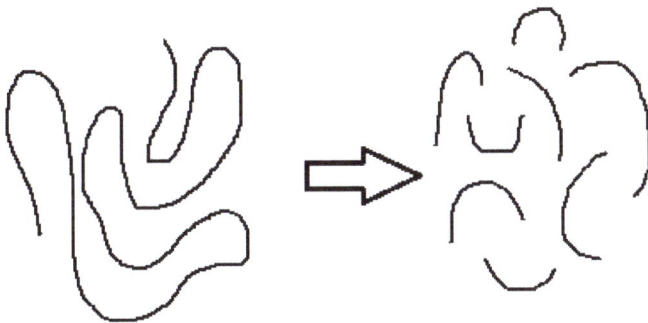

Acid treatment of starch to reduce the molecular chain length

Polyvinyl Alcohol (PVA)

Polyvinyl alcohol (PVA) is a very versatile sizing material. It can be used for sizing cotton, rayon, polyester and their blends. It is manufactured by polymerizing vinyl acetate monomers and then substituting the acetate groups with hydroxyl groups by hydrolysis (figure below). The properties of the PVA are largely governed by the degree of substitution.

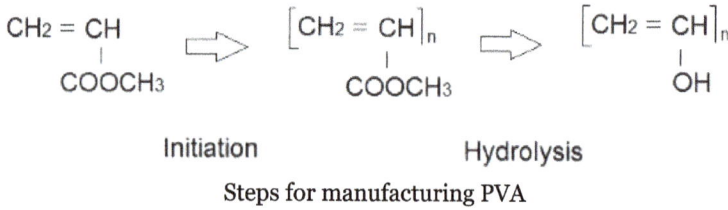

$$CH_2 = CH \quad \Rightarrow \quad \left[CH_2 = CH \right]_n \quad \Rightarrow \quad \left[CH_2 = CH \right]_n$$
$$| \qquad\qquad\qquad | \qquad\qquad\qquad\qquad |$$
$$COOCH_3 \qquad\qquad COOCH_3 \qquad\qquad\qquad OH$$

Initiation   Hydrolysis

Steps for manufacturing PVA

If the PVA is hydrolyzed to the maximum possible extent (>99%), then the formation of hydrogen bonding becomes very intense and thus the strength of the PVA film becomes very high (figures below). Besides, its solubility in water is lower as compared to partially hydrolyzed PVA. Thus the desizing becomes difficult and higher temperature is required for desizing. Therefore, this grade (super hydrolyzed) is generally not preferred for sizing.

Partially hydrolyzed PVA exhibits lower film strength. The presence of a big functional group in the side chain reduces the strength of the partially hydrolyzed PVA film. Thus it provides the advantages of easy splitting (separation) of yarns after drying and less disruption of size film. Besides, partially hydrolyzed PVA shows better adhesion with the hydrophobic fibres than the fully hydrolyzed PVA (following table).

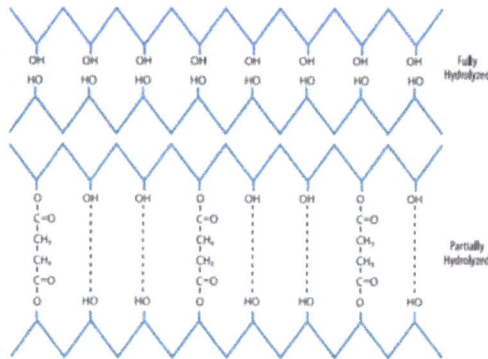

Hydrogen bonding in PVA

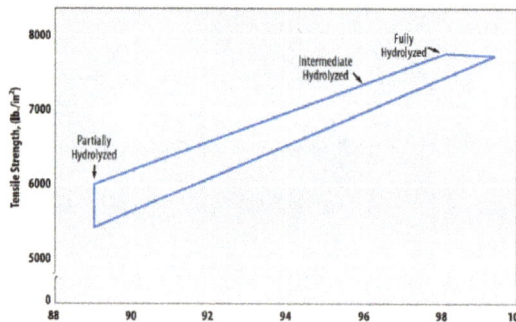

Effect of hydrolysis on the strength of PVA film.

Solubility in water is better for the partially hydrolyzed PVA. In partially hydrolyzed PVA, the acetate group, due to its bigger size, acts as an obstacle against the close packing of molecular chain (figure below). Therefore, the penetration of water molecule is rather easy in case of partially hydrolyzed PVA making the desizing operation facile.

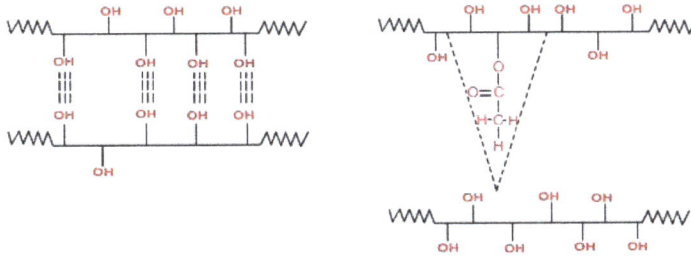

Molecular chains of fully hydrolyzed and partially hydrolysed PVA

Table: Adhesion of PVA with synthetic fibres

Type of fibre	Adhesion strenght (g/mm2)	
	Partially hydrolyzed	Fully hydrolyzed
Acetate	10	2
Nylon	11	6
Acrylic	9	4
Polyester	7	1

Figure below depicts the abrasion resistance performance of yarns sized with partially, intermediate and fully hydrolyzed PVA. The former provides better waving performance in terms of the following.

• Less shedding or dropping of size

• Lower yarn hairiness

• Lower size add-on

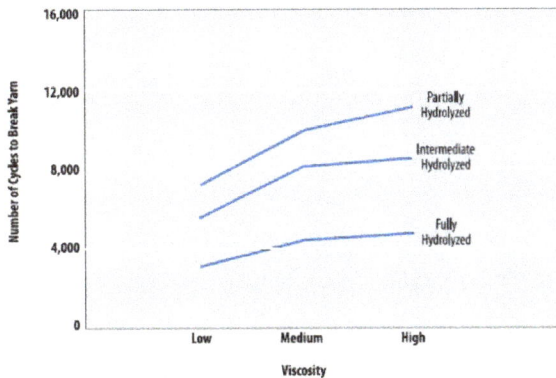

Abrasion resistance of yarns sized with PVA

Table: Degree of hydrolysis in different grades of PVA

PVA grade	Degree of hydro-lysis	Application
Super hydrolysed	>99%	Not a preferred material for sizing
Fully hydrolysed	98-99%	100% Cotton
Intermediate hydrolysed	95-98 %	Polyester and other synthetic fibres and blends
Partially hydrolysed	87-90	Polyester and other synthetic fibres and blends

PVA is available in different viscosity range based on its degree of polymerization (DP). Higher DP leads to greater molecular weight and viscosity. The viscosity ranges from 5 cP to 65 cP. For most of the spun yarns, medium viscosity is preferred. If the viscosity is very low then size paste penetrates too much within the yarn structure affecting the flexibility and extensibility. On the other hand, if the viscosity is too high, then size film forms over the yarn surface without much adhesion with the fibres. These surface coatings of size materials are prone to dropping. To maintain the balance, a mixture of low viscosity and high viscosity PVA is often used.

Typical Recipe of Sizing

Carded Cotton Yarn

Modified Starch   : 10.5 % on the weight of water
PVA   : 2.86 %
Acrylic binder (liquid) : 6.6 %
Lubricant   : 0.7 %.

Combed Cotton Yarn

Modified Starch   : 12.5 % on the weight of water
PVA   : 3.0 %
Acrylic binder (liquid) : 6.91 %
Lubricant   : 0.87 %.

Polyester-cotton Blended Yarn

Modified Starch   : 13 % on the weight of water
PVA   : 3.6 %
Acrylic binder (liquid) : 8.4 %
Lubricant   : 0.87 %.
Antistatic   : 0.5 kg

Typical Steps Followed in the Industry to Prepare the Size Paste

- Take standard volume (700 litres) of water (normal temperature) into the pre-mixture through water flow meter, start the stirrer.

- Properly weigh all the chemicals required.

- Add modified starch, PVA and then Acrylic binder slowly into the pre-mixture vessel.

- Stir the mixture for 15 minutes.

- Start the stirrer of the cooker, transfer the mixture to the closed cooker from pre-mixture vessel. Close all the open valves of the cooker.

- Heat the cooker mixture with the injection of direct steam, steam will auto cut when the cooker temperature reaches the preset limit (110° C). It takes around 30 to 35 minutes.

- Cook the mixture for another 40 minutes, the temperature of the cooker will reach to around 120 to 125° C depending on the size recipe.

- Start the stirrer of the storage vessel and then transfer the cooked size paste to the respective storage vessels.

- Add lubricant with the size paste at storage.

- Maintain storage temperature at around 80-90° C.

- Paste is ready to transfer to the sow box of sizing machine.

Sizing Machine

The sizing machine can be divided into four main zones as shown in the figure. The zones are

- Creel zone

- Size box zone

- Drying zone

- Headstock zone

The creel zone contains large number of warper's beam which can be arranged in different fashion depending on the design of the creel. Individual warp sheet emerging from warper's beam are merged together to form the final warp sheet which passes through the size box. During the passage through the size box, the warp sheet picks up size paste and holds a part of the paste after squeezing. Then the wet warp sheet passes through the drying zone and wound on the weavers beam.

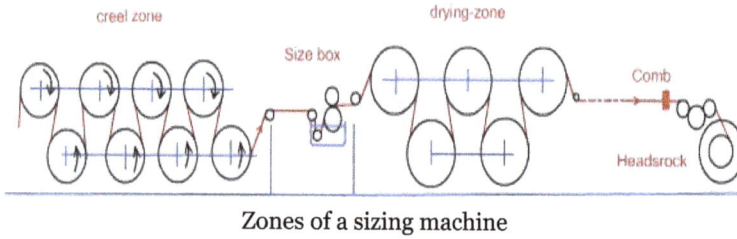

Zones of a sizing machine

Creel Zone

The creel zone of a sizing machine can have following types of design:

- Over and under creel
- Equi-tension creel
- Vertical creel
- Inclined creel

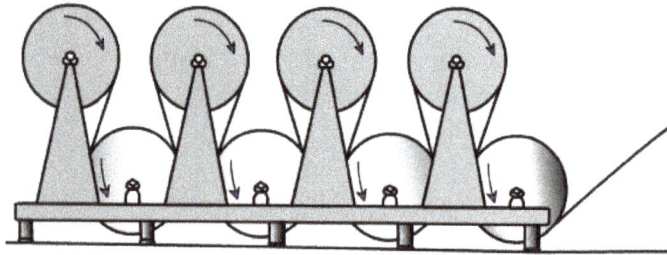

Over and under creel

In case of over and under creel, the warper's beams are arranged in two rows, having different heights, in an alternate manner. The warp sheet coming out from the rearmost beam passes under the second beam and over the third beam and so on. The individual warp sheets coming out from beams are merged together to form the final warp sheet. The warp sheet coming from the rearmost beam definitely experiences more tension and stretch than the warp sheet coming from the beam located nearest to the size box. The problem is partially mitigated when two creels are used one for each of the two size boxes as shown in the following figure. If there are twelve beams then six beams are mounted on creel one and remaining six beams are mounted on creel two reducing the over and under movement of the warp sheet.

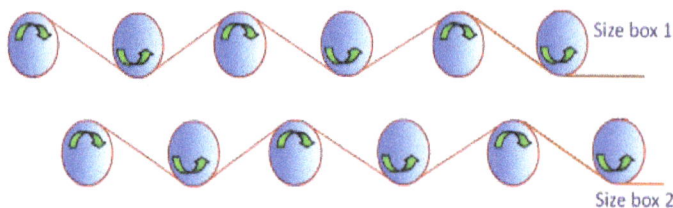

Over and under creel for two size boxes

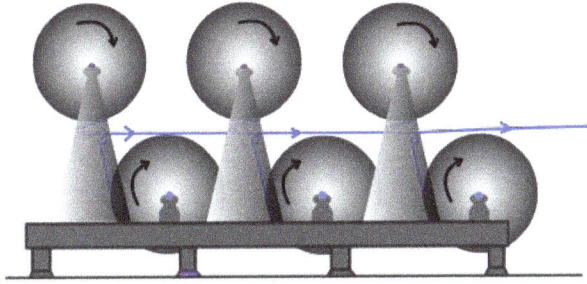

Equi-tension creel

In case of equi-tension creel the pattern of movement of warp sheet is completely different than that of over and under creel. In equi-tension creel, warp sheet does not move over and under any beam. One small guide rollers provided with every beam which deflects the warp sheet towards the proper path. Here, the warp sheets are subjected to equal tension and stretch irrespective of the position of the warper's beam.

Another improvement in this direction has been implemented in the inclined creel. Here the height of the beam changes based on its position so that a constant inclination can be maintained in the path of the warp sheet.

All the designs, which have been discussed till now, requires considerable amount of floor space. This can be solved if vertical creels are used. In vertical creels, the beams are stacked vertically as shown in the following figure.

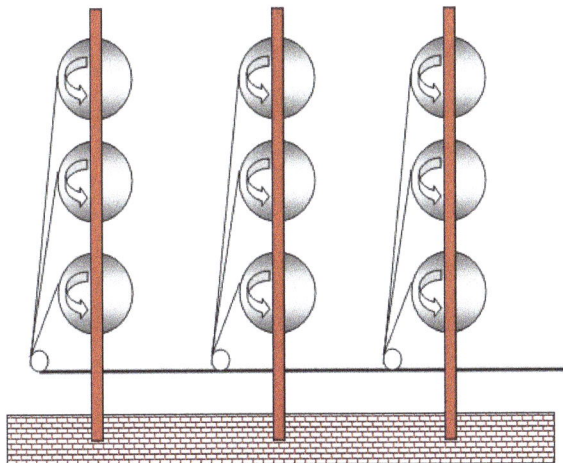

Vertical creel

It is very important to maintain adequate and uniform tension in the warp sheet during the entire sizing process. However, as the sizing process continues, the radius of the warper's beam reduces. Therefore, it is required to adjust the warp tension by adjusting either the dead weight suspended with the rope passing over the ruffles of the warper's beam (figure) or by controlling the pneumatic pressure applied on the bearing region of warper's beam.

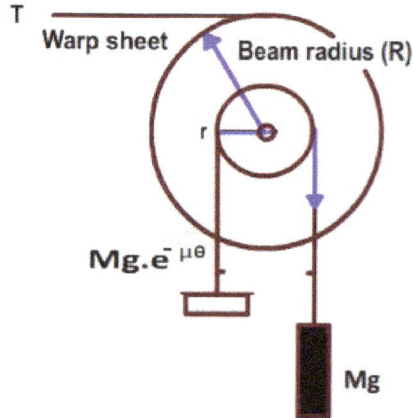

Warp tension control by dead weight system

From the balance of moment the following equation can be formulated.

$$T.R = 2T_2 \times r\left(1 - e^{-\mu\theta}\right)$$

$$or\ T.R = 2 \times Mg \times r\left(1 - e^{-\mu\theta}\right)$$

where T is the warp tension, R is the radius of warper's beam, T_2 is the tension in the tight side of the rope, r is the ruffle radius, μ is the coefficient of friction between the rope and the ruffle, θ is the angle of wrap (in radian) of the rope over the ruffle, M is the mass of suspended element and g is acceleration due to gravity.

In sizing process, the allowable stretch is 1-1.5% for cotton and polyester yarns. The stretch can be higher (3-5%) for viscose and acrylic yarns.

Size Box Zone

This is the zone where the warp sheet is immersed into the size paste and then squeezed under high pressure so that uniform coating of size film forms over the yarn surface. The process of immersion is called 'dip' and the process of squeezing by means of a pair of squeezing rollers is called 'nip'. The size box can have different number combinations of 'dip' and 'nip' to meet the requirement of various yarns. For filament yarns 'one dip and one nip' is preferred where as for spun yarns made from staple fibres 'two dip and two nip' is advisable.

Two dip and two nip process allows grater time for immersion of yarns within the size paste and thus this process forms more uniform coating of size film. When the yarns are squeezed by the first pair of squeeze rollers, yarns become compressed. When the yarns come out of the nip of squeezing rollers, they try to regain their original arrangement and therefore an inward pressure is created which causes more penetration of size materials within the yarn structure.

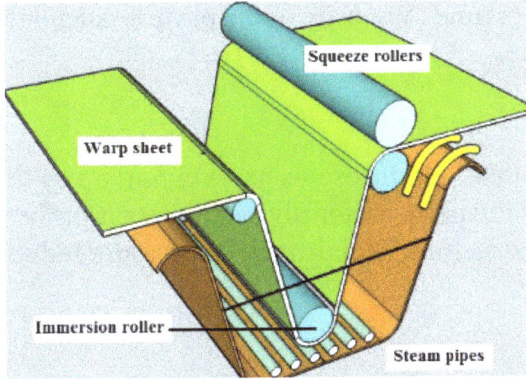

One dip one nip size box

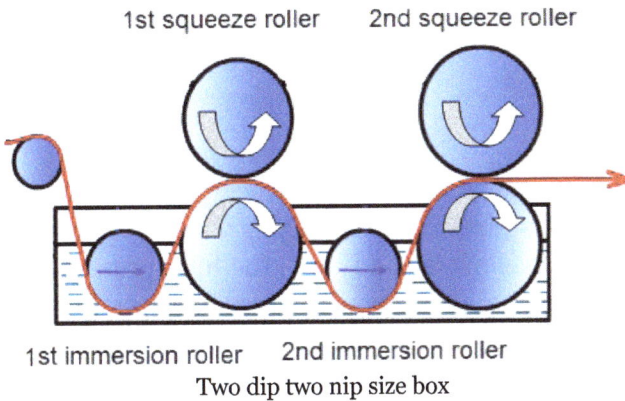

1st squeeze roller   2nd squeeze roller

1st immersion roller   2nd immersion roller

Two dip two nip size box

Important Parameters

The wet pick-up by the warp sheet is influenced by the following parameters:

i. Viscosity of Size Paste

Viscosity of a fluid indicates its resistance against the flow. The viscosity of the size paste is mainly influenced by the concentration (solid content) and temperature of size paste. Higher concentration implies higher viscosity. Viscosity of size paste reduces with the increase of temperature. The wet pick-up generally increases with the increase in viscosity. Viscosity also determines the penetration of size paste within the yarn structure. If more penetration is desired then viscosity should be lowered and vise versa. For bulky yarns, penetration is relatively easy and therefore higher viscosity may be preferred.

Viscosity of size paste can be measured by Zahn cup. It is a stainless steel cup with a small hole at the centre of the bottom of the cup. A long handle attached to the sides of the cup. There are five cup specifications, labelled Zahn cup #N, where N is the number from one through five. Large number cup sizes are used when viscosity is high, while low number cup sizes are used when viscosity is low. To determine the viscosity, the cup is dipped and completely filled with the size paste. After lifting the cup out of the paste the user measures the time until the paste streaming out of it breaks up, this is the corresponding 'efflux time'. Viscosity of the paste is calculated from the efflux time using standard formulae.

ii. Squeezing Pressure

The squeeze pressure forces out the excess paste picked up by the warp sheet. Besides, the pressure distributes the paste uniformly over the yarn surface and causes size penetration within the yarn structure. Higher squeeze pressure reduces the wet pick-up and add-on% (figure below).

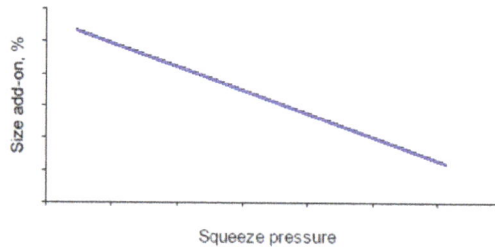

Effect of squeeze pressure on size add-on %

The effect of high pressure squeezing during sizing was investigated by Hari et al. It was found that for the same level of size add-on%, the high pressure squeezing facilitates better penetration of size within the yarn structure. However, the thickness of coating outside the yarn periphery reduces at high pressure squeezing. This reduces the dropping of size during weaving. The comparison of the size coating and penetration at high and low squeezing pressure is presented in the table.

Table: Comparison between high and low pressure squeezing

Pressure	Size coating (film thickness)	Size penetration
High	Low	High
Low	High	Low

Though there was no significant difference in the tensile properties of yarns sizes using high and low pressure, the weaveability of the former was much better than the latter.

iii. Hardness of Top Squeeze Roll

The bottom squeezing roller is made up of stainless steel. The top squeezing roller is having a metallic core part which is covered with synthetic material. If the hardness of the top roller is low, then there will be flattening of the roller. Thus the contact area increases which effectively reduces the pressure acting at the nip zone. Therefore, the size pick-up increases. In contrast, harder rollers give sharper nip and lower wet pick-up. The shore hardness of the top roller is around 45°.

iv. Thickness of Synthetic Rubber on the Top Roller.

If the thickness of synthetic rubber cover on the top roller is greater, then the extent of flattening is more. This will reduce the nip pressure and thus the wet pick-up will increase.

v. Position of Immersion Roller

The position of the immersion roller within the size box is adjustable. If the height of immersion roller is lowered then the residence time of the warp sheet within the size paste increases. This will lead to the increase in wet pick-up if other factors are constant.

vi. Speed of Sizing

Speed of sizing also influences the wet pick-up by the warp sheet.

- Higher speed reduces the residence time of the yarn within the paste which should reduce the wet pick-up.

- Higher speed increases the drag force between the warp sheet and size paste which should induce more flow of paste with the warp sheet.

- Higher speed reduces the time of squeezing which should increase the wet pick-up.

The speed of sizing will influence the wet pick-up based on the preponderance of the aforesaid factors. In modern sizing machine, the practical speed can be around 100 m/min. Though machine manufactures claim that the speed can be as high as 150 m/min.

Sizing Diagram

Figure below presents the schematic relationship between the concentration of size paste and final add-on% of size on the yarn.

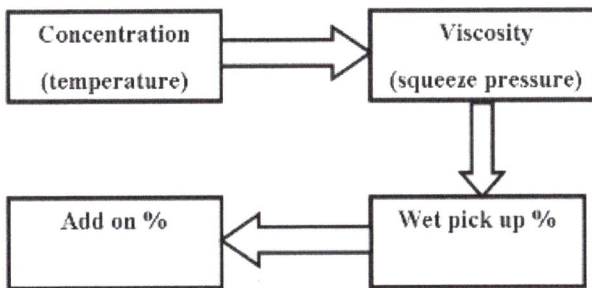

Relationship between concentration and add-on%

- At a given temperature, concentration of size paste determines the viscosity.

- For a given yarn, squeeze pressure and sizing speed, viscosity of size paste determines the wet pick-up.

- Wet pick-up determines the add-on. For a given concentration of size paste, higher wet pick-up leads to higher add-on and vice versa.

- Therefore, if other conditions are same, then there will be a functional relationship between size paste concentration and add-on.

Figure below depicts the relationship among concentration, viscosity, wet pick-up and add-on. Thin boiling starch requires higher level of concentration, than the normal starch, for the same level of viscosity. Therefore, even if the wet pick-up is same, add-on will be higher (due to higher concentration) for thin boiling starch. Besides, for the same level of add-on, wet pick-up will be less for thin boiling starch. Therefore, water evaporation during drying will be lower for thin boiling starch as compared to normal starch. This will lead to energy saving in case of sizing with thin boiling starch.

From the figure, it can be seen that thin boiling starch requires higher concentration (C_2) than noemal starch (C_1) for creating the same level of viscosity (η_1). This will lead to same wet pick-up for both types of starch. Now, add-on will be higher for the thin boiling starch (ΔA) even at the same level of wet pick-up (W_1). Thus, in the third quadrant, the broken line representing the thin boiling starch is positioned below the solid line representing normal starch.presents the schematic relationship.

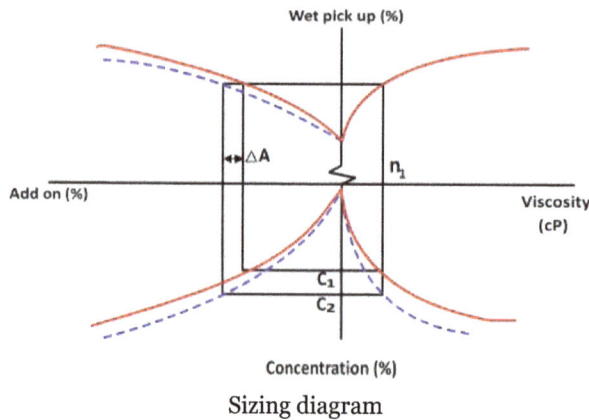

Sizing diagram

The modern sizing practice recommends the use of high concentration (which results in high viscosity) of size paste and high squeezing pressure.

Different combinations of squeezing pressure and concentration

- High pressure and high viscosity combination is preferred as high pressure reduces the wet pick-up and high concentration ensures that targeted add-on will be attained with minimum water evaporation.

- Hypothetically, low pressure and low viscosity combination can also give same level of wet pick-up as obtained with high pressure and high viscosity combination. But the same level of add-on cannot be obtained after drying due to low concentration.

- Low concentration and high pressure will give the minimum wet pick-up.

Example: The targeted add-on is 10 % and oven dry mass of supply warp sheet is 100 kg. If the concentration is 20% then high pressure can be used to achieve wet pick-up of 50 kg. Then in the drying section 40 kg water will be evaporated to get the target add-on of 10%. In contrast, if the concentration is 10% then low pressure can be used to achieve wet pick-up of 100 kg. Then in the drying section 90 kg water will be evaporated to get the targeted add-on of 10%. This obviously requires more energy consumption during drying.

Crowning of Top Roller

High pressure squeezing is used to reduce the load on the drying system. In modern sizing machine, the squeezing force can go up to the level of 100,000 N. This force is applied on the two sides of the metallic core of top squeeze roller. This pressure is good enough to cause bending in the top squeeze roller which may result uneven pressure along the nip line. To overcome this problem, crowned top rollers are used. The synthetic rubber coated surface of the top squeeze roller is subjected to grinding operation so that the diameter at the sides is lower as compared to that of at the middle as shown in the figure. This is compensated by the bending of the top rollers and uniform pressure is obtained along the nip line.

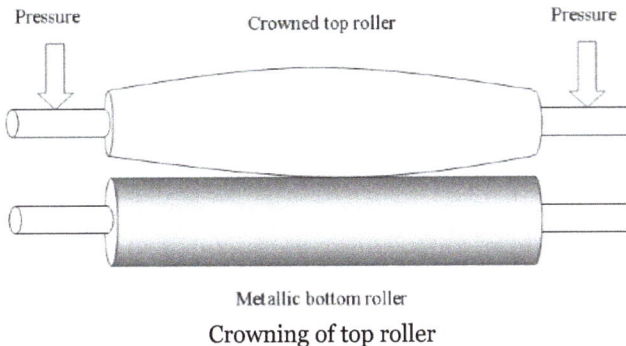

Crowning of top roller

Percent Occupation and Equivalent Yarn Diameter

The relative closeness of the yarns inside the size box is expressed by percentage occupation and equivalent yarn diameter. 100% occupation means that yarns are physically touching each other. The number of yarns with 100% occupation can be calculated approximately if nominal yarn diameter is known. Equivalent yarn diameter indicates the space between the two yarns in terms of yarn diameter. If equivalent yarn diameter is zero that means that the yarns are touching each other i.e. 100% occupation. Figure depicts the situation with 100% occupation i.e. zero equivalent yarn diameter. Figure presents the situation with 50% occupation i.e. one equivalent yarn diameter.

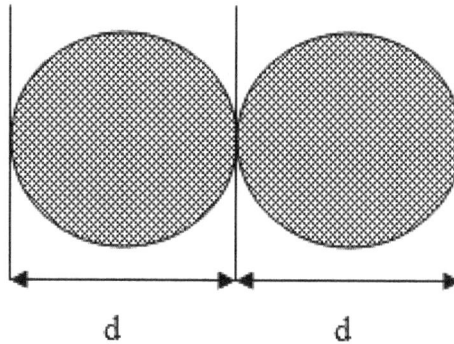

d   d

Yarn arrangement for 100% occupation

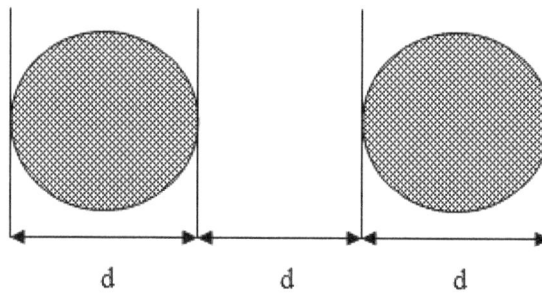

d   d   d

Yarn arrangement for 50% occupation

$Percent\ occupation$

$$= \frac{Actual\ number\ of\ yarns\ in\ the\ warp\ sheet}{Number\ of\ yarns\ in\ the\ warp\ sheet\ with\ 100\%\ occupation}$$

The percent occupation and equivalent yarn diameter are related with the following expression.

$Percent\ occupation$

$$= \frac{100}{1 + Equivalent\ yarn\ diameter}$$

If the percent occupation is very high then the yarn may not be uniformly coated by the size film. For warp sheet having very large number of yarns, it may be preferable to use two size boxes to keep the percent occupation value within permissible range.

Drying Zone

This is the zone where the wet yarns are dried by evaporating the water from the size paste. The drying operation is very crucial because of the following reasons.

1. It consumes most of the energy of sizing process

2. Inadequate drying will cause sticking of yarns with one another causing problem in weaving

3. Over-drying will make the size film brittle and therefore they may fall apart by minimum abrasion.

Drying is done by passing the warp sheet over large number of drying cylinders, coated with Teflon (poly tetra fluro-ethylene), arranged in sequential manner. The number of drying cylinders can vary from 2 to 30 depending on the amount of water to be evaporated in unit time. In general, higher speed of sizing would require more number of drying cylinders. The following expressions are useful for calculating the mass of water to be evaporated during drying.

$Mass\,(kg)\,of\,water\,to\,be\,evaporated\;per\,unit\,oven\,dry\,mass\,(kg)\,of\;yarn$

$$= \left(\frac{Add\;on\,\%}{Concentration\,\%} \right) - \left(\frac{Add\;on\,\%}{100} \right)$$

The first part of the expression yields wet pick-up. If the mass of dry size is subtracted from the wet pick-up then the amount of water to be evaporated can be obtained. The above equation presumes that there is no residual moisture in the sized yarn after drying. However, for the running machine it is more important to calculate the mass of water to be evaporated in unit time (minute). This will be depending on the following factors.

- Sizing machine speed

- Total number of yarns

- Linear density of yarns (tex)

- Add-on %

- Concentration %

The mass of yarn passing through the machine per minute can be expressed as follows:

$$= \frac{Sizing\;machine\;speed\,(m\,/\,min) \times Total\;number\;of\;yarns \times tex}{1000 \times 1000}\,kg$$

The mass of paste picked up by the warp sheet per minute will be

$$= \left(\frac{Sizing\ machine\ speed\ (m/\min) \times Total\ number\ of\ yarns \times tex}{1000 \times 1000} \times wet\ pick\ up \right) kg$$

$$= \left(\frac{Sizing\ machine\ speed\ (m/\min) \times Total\ number\ of\ yarns \times tex}{1000 \times 1000} \times \frac{add\ on\ \%}{concentration\ \%} \right) kg$$

The mass of water to be evaporated per minute *will be*

$$= \frac{Sizing\ machine\ speed\ (m/\min) \times Total\ number\ of\ yarns \times tex}{1000 \times 1000}$$

$$\times \frac{add\ on\ \%}{concentration\ \%} \times \left(1 - \frac{Concentration\ \%}{100} \right) kg$$

The result obtained from above expression has been depicted in the following figure. For the same level of size add-on, more water has to be evaporated if the size paste concentration is low.

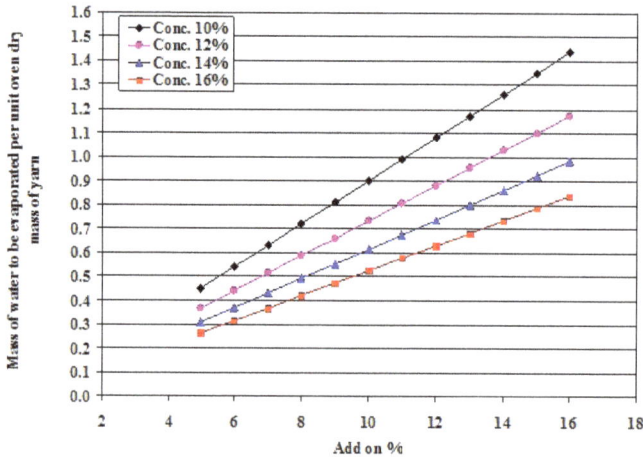

Effect of add-on% and concentration % on the amount of water evaporation

Methods of Drying

The methods of drying in sizing process can broadly be divided in two categories.

- Conduction method

- Convection method

In conduction method the warp sheet is passed over a metallic cylinder which is heated by using superheated steam. Heat exchange takes place between the wet warp sheet and heated cylinders and in the process the warp sheet is dried (figure a). The efficiency of this process is very high. The problem of this system is that only one side of the warp sheet is exposed to the heated cylinder at a time. This problem can be overcome in convection method. In convection method, hot air is circulated within an enclosed chamber and the warp sheet passes through the chamber with the help of some guides (figure b). Both the sides of the warp sheet is exposed to the hot air at the same which ensures that the drying is very uniform. However, the efficiency of the process is lower as compared to that of conduction process.

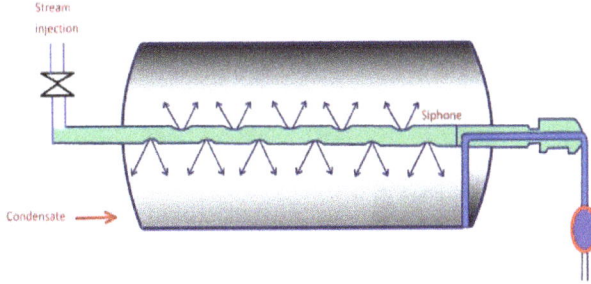

(a) Conduction drying method

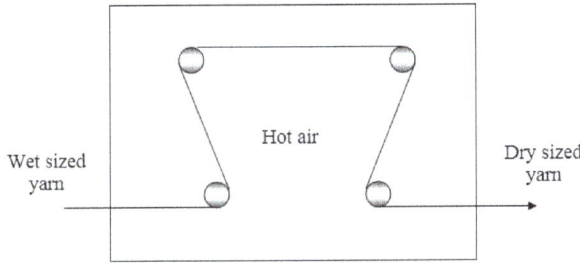

(b) Convection drying

Design of the drying zone can be of various types. To ensure better drying and reduction of load on individual cylinders, the wet warp sheet is often splitted into multiple sheets. Each sheet is then dried by a separate group of drying cylinders (figure c). This initial drying is generally done at relatively lower temperature. Finally all the sheets are again merged and final drying takes place using another set of drying rollers. The temperature range for drying of cotton warp is 100-140°C.

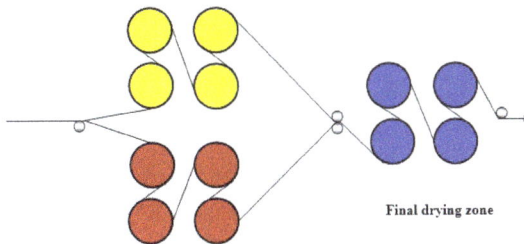

(c) Two zone drying

Splitting

After drying, the warp sheet is splitted so that the yarns regain their individual identity before they are wound on the weaver's beam. This is depicted in the figure.

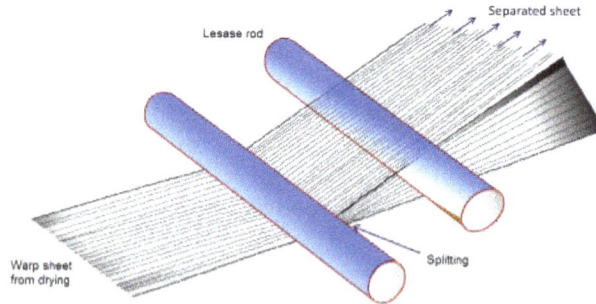

Splitting of warp sheet

Splitting is required because warp sheet coming out of the drying section adhere to each other depending on the efficiency of the pre-drying section. Lease rods which are often coated with chromium are used to split the warp sheet in a systematic manner as shown in figures below. Figure above depicts a situation when each of the lease rods split the incoming warp sheet into two equal parts. On the other hand, Figure presents a situation where warp sheet coming from a warper's beam is separated at a time by a lease rod. So, if it is assumed that warp sheet is originating from four warper's beams placed on the creel, the first lease rod will split the warp sheet into two unequal parts. One part will have yarns from a particular warper's beam whereas the other part will consist of yarns form the remaining three warper's beam.

Function of lease rods is to separate the individual yarns which are stuck together by dried size. During the splitting some amount of size film would be dropped as waste. However, a large number of longer fibres, bridging two adjacent yarns would also get broken into smaller pieces. Therefore, splitting is considered to have some beneficial effect from hairiness viewpoint. This has been represented pictorially in the last figure.

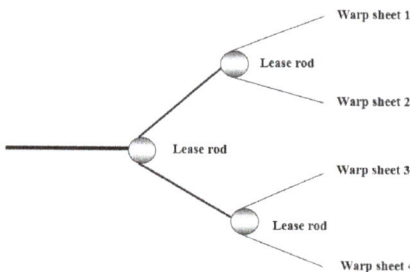

Splitting of sized warp sheet into equal parts   Splitting of sized warp sheet into unequal parts

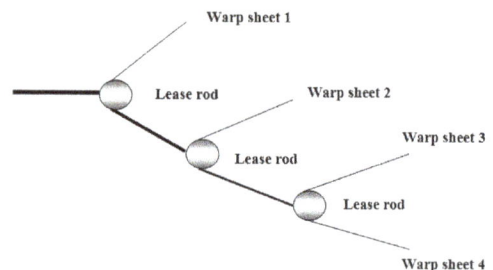

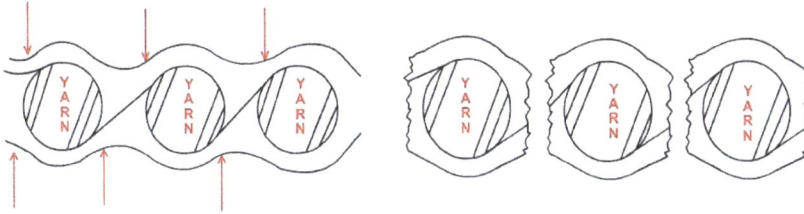

Arrangement of hairs in yarns before and after splitting.

Beaming

After the splitting, the warp sheet is finally wound on the weaver's beam. The warp sheet passes through an adjustable reed which can be expanded or collapsed based on the width of the beam.

Pre-wetting of Yarns before Sizing

Pre-wetting is done for the staple spun yarns to make the sizing process more efficient. The warp sheet is passed through a box which contains hot water (temperature around 90°C) and thus the waxes and other impurities are partially removed. This improves the adhesion between the yarns and the size materials. If the sizing machine is having two size boxes then the first one can be used for the pre-wetting and the second one for sizing. Generally, size boxes having two dip and two nip are preferred when pre-wetting is done. After pre-wetting, water occupies the core of the yarns and thus the penetration of the size within the yarn structure reduces and uniform size film is formed over the yarn surface. High squeezing pressure is used at the nip of pre-wetting box so that the water retained by the yarns is minimized. This precludes the possibility of dilution of paste concentration in the size box as well as reduction of paste temperature. The advantages of pre-wetting are as follows:

- Reduction of size ingredient consumption up to 50%

- Increase in yarn strength

- Reduction of yarn hairiness

- Improvement of loom efficiency

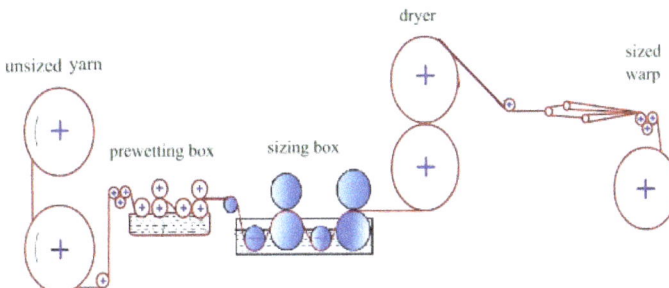

Simplified representation of pre-wetting process

Quality Evaluation of Sized Yarns

The real performance of sized yarn can only be appraised during the weaving operation. However, the performance of sized yarns can be forecasted by judging the following quality parameters:

- Tenacity and breaking elongation of sized yarn
- Cohesiveness and adhesion of the sized film
- Abrasion resistance
- Fatigue resistance

Tensile strength (tenacity) enhancement after sizing does not exhibit good correlation with the actual warp breakage rate during weaving as tension acting on yarn does not exceed 20% of the yarn breaking strength. Cohesiveness of size film is evaluated by measuring the tensile properties of a thin film made from the size paste. A comparison of cohesive properties of maize starch, PVA and CMC films is given in the table.

Table: Properties of starch, CMC and PVA

Mechanical properties	Size typ		
	Starch	CMC	PVA
Tenacity (cN/mm²)	3.52	3.33	4.03
Elongation %	20.56	30.29	45.24
Initial modulus, (cN/mm²)	200.87	150.24	90.80

PVA demonstrates higher cohesiveness and elongation than Maize starch and CMC. Adhesive power indicates the compatibility of the size material with the textile substrates.

$$Adhesive\ power = \frac{Breaking\ strength\ of\ sized\ roving\ at\ gauge\ length\ `l'}{Breaking\ strength\ of\ sized\ roving\ at\ zero\ gauge\ length}$$

`l' is greater than staple length of fibre

If the adhesion between the fibres and size film is good then the slippage of the fibres in the sized roving during tensile testing will reduce. This will increase the adhesive power. When adhesion is good, the critical adhesive power is attained at lower add-on as shown in the figure with solid line. In contrast, if the adhesion power is bad, then higher add-on is required to reach the level of critical adhesive power.

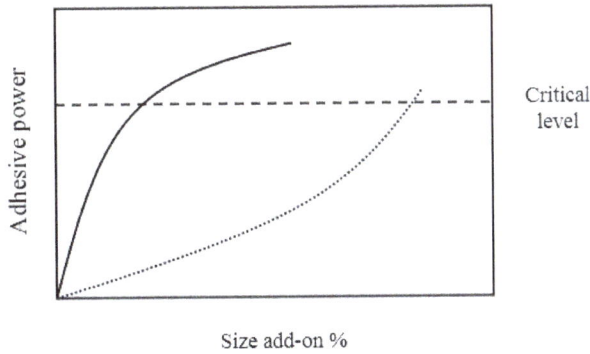

Size add-on % vs adhesive power

During weaving, abrasion takes place between warp yarns and loom parts like heald eyes, reed and shuttle. Sized yarns can be subjected to abrasion tests and the number of cycles required to break a given number of yarns is noted. These observations can be used to calculate the mean abrasion cycle that can be resisted by a sized yarn. Alternatively, sized yarns can be subjected to a fixed number of abrasion cycles and then the % deterioration in terms of tensile strength is calculated using the following expression. Lower deterioration implies good quality of sizing and vice versa.

$$\% \ Deterioration \ due \ to \ abrasion$$
$$= \frac{Original \ breaking \ strength \ of \ yarn - breaking \ strength \ of \ yarn \ after \ abrasion}{Original \ breaking \ strength \ of \ yarn} \times 100$$

During the weaving process, the yarn undergoes repeated extension and bending. This causes cumulative damage to the fibre and yarn structure. As a result the yarn actually fails, due to fatigue, at a breaking load which is much lower than its actual breaking load. Even the very strong metal wires break after repeated flexing due to poor fatigue resistance. Sulzer-Ruti Webtester is used to evaluate the fatigue resistance of sized yarns. The instrument simulates the actual weaving conditions.

Sizing

Sizing or size is any one of numerous substances that is applied to, or incorporated into, other materials especially papers and textiles to act as a protective filler or glaze. Sizing is used in papermaking and textile manufacturing to change the absorption and wear characteristics of those materials.

Sizing is used for oil-based surface preparation for gilding (also known as *mordant* in this context). It is used by painters and artists to prepare paper and textile surfaces for some art techniques.

The term sizing also refers to the process of including or applying the substance.

Papermaking

Sizing is used during paper manufacture to reduce the paper's tendency when dry to absorb liquid, with the goal of allowing inks and paints to remain on the surface of the paper and to dry there, rather than be absorbed into the paper. This provides a more consistent, economical, and precise printing, painting, and writing surface. This is achieved by curbing the paper fibers' tendency to absorb liquids by capillary action. In addition, sizing affects abrasiveness, creasibility, finish, printability, smoothness, and surface bond strength and decreases surface porosity and fuzzing.

There are three categories of papers with respect to sizing: *unsized* (*water-leaf*), *weak sized* (*slack sized*), and *strong sized* (*hard sized*). Waterleaf has low water resistance and includes absorbent papers for blotting. Slack sized paper is somewhat absorbent and includes newsprint, while hard sized papers have the highest water resistance, such as coated fine papers and liquid packaging board.

There are two types of sizing: *internal sizing,* sometimes also called engine sizing, and *surface sizing* (tub sizing). Internal sizing is applied to almost all papers and especially to all those that are machine made, while surface sizing is added for the highest grade bond, ledger, and writing papers.

Surface Sizing

Surface sizing solutions consists of mainly modified starches and sometimes other hydrocolloids, such as gelatine, or surface sizing agents such as acrylic co-polymers. Surface sizing agents are amphiphilic molecules, having both hydrophilic (water-loving) and hydrophobic (water-repelling) ends. The sizing agent adheres to substrate fibers and forms a film, with the hydrophilic tail facing the fiber and the hydrophobic tail facing outwards, resulting in a smooth finish that tends to be water-repellent. Sizing improves the surface strength, printability, and water resistance of the paper or material to which it is applied. In the sizing solution, optical brightening agents (OBA) may also be added to improve the opacity and whiteness of the paper or material surface.

Internal Sizing

Internal sizing chemicals used in papermaking at the wet end are alkyl succinic anhydride (ASA), alkyl ketene dimer (AKD) and rosin. By making the paper web more hydrophobic, the sizing agents influence dewatering and retention of fillers and fibers in the paper sheet. Next to paper quality, internal sizing agents' main effect is on runability of the paper machine.

Preservation

While sizing is intended to make paper more suitable for printing, it also makes printing paper less durable and poses a problem for preservation of printed documents. Sizing with starch was introduced quite early in the history of papermaking Dard Hunter in *Papermaking through Eighteen Centuries* corroborates this by writing, "The Chinese used starch as a size for paper as early as A.D. 768 and its use continued until the fourteenth century when animal glue was substituted." In the early modern paper mills in Europe, which produced paper for printing and other uses, the sizing agent of choice was gelatin, as Susan Swartzburg writes in *Preserving Library Materials'*: "Various substances have been used for sizing through the ages, from gypsum to animal gelatin." Hunter describes the process of sizing in these paper mills in the following:

> The drying completed, the old papermakers dipped their paper into an animal size that had been made from the parings of hides, which they procured from the parchment-makers. It was necessary to size that paper so that it would be impervious to ink, but sizing was more needed in writing than in printing papers. Many books of the fifteenth century were printed upon paper that had not been sized, this extra treatment not being essential for a type impression. The sizing was accomplished by a worker holding a number of sheets by the aid of two wooden sticks, and dipping the paper into the warm gelatinous liquid. The sheets were then pressed to extract the superfluous gelatine. This crude method of sizing the paper was extremely wasteful as many sheets were torn and bruised beyond use. The sizing room of the early paper mills, was, for this reason, known as the slaughter-house.

With the advent of the mass production of paper, the type of size used for paper production also changed. As Swartzburg writes, "By 1850 rosin size had come into use. Unfortunately, it produces a chemical action that hastens the decomposition of even the finest papers." In the field of library preservation it is known "that acid hydrolysis of cellulose and related carbo-hydrates is one of the key factors responsible for the degradation of paper during ageing." Some recent professional work has focused on the specific in the degradation involved in the deterioration of paper that has had a rosin sizing process, and what amount of rosin affects the deterioration process, in addition to work on developing permanent paper and sizing agents that will not eventually destroy the paper. An issue on the periphery to the preservation of paper and sizing, is washing, which is described by V. Daniels and J. Kosek as, "The removal of discolouration ... in water is principally effected by the dissolution of water-soluble material; this is usually done by immersing paper in water." In such a process, surface level items applied to the paper, such as size in early paper making processes as seen above, have the possibility of being removed from the paper, which might have some item specific interest in a special collections library. With later processes in paper making being more akin to "engine sizing," as H. Hardman and E. J. Cole describe it,

"Engine sizing, which is part of the manufacturing process, has the ingredients added to the furnish or stock prior to sheet formation," the concern for the removal of size is less, and as such, most literature focuses on the more pressing issue of preserving acidic papers and similar issues.

Gilding

Sizing is a term used for any substance which is applied to a surface before gilding in order to ensure adhesion of the thin gold leaf to the substrate. Egg whites have often been used as sizing; the Ancient Egyptians sometimes used blood. Other commonly used traditional materials for gold leaf sizing are rabbit skin glue diluted and heated in water (water gilding), and boiled linseed oil (oil gilding); modern materials include polyvinyl acetate.

Textile Warp Sizing

Sizing of the warp yarn is essential to reduce breakage of the yarn and thus production stops on the weaving machine. On the weaving machine, the warp yarns are subjected to several types of actions i.e. cyclic strain, flexing, abrasion at various loom parts and inter yarn friction.

Sizing the warp

With sizing, the strength — abrasion resistance — of the yarn will improve and the hairiness of yarn will decrease. The degree of improvement of strength depends on adhesion force between fiber and size, size penetration as well as encapsulation of yarn. Different types of water soluble polymers called textile sizing agents/chemicals such as modified starch, polyvinyl alcohol (PVA), carboxymethyl cellulose (CMC), acrylates are used to protect the yarn. Also wax is added to reduce the abrasiveness of the warp yarns. The type of yarn material (e.g. cotton, polyester, linen), the thickness of the yarn, type of weaving machinery will determine the sizing recipe.

The sizing liquor is applied on warp yarn with a warp sizing machine. After the weaving process the fabric is desized (washed).

Sizing may be done by hand, or in a sizing machine.

References

- Altaf H. Basta and others, "The Role of Neutral Rosin-Alum Size in the Production of Permanent Paper." Restaurator: International Journal for the Preservation of Library and Archival Material, 27, no. 2 (2006): 67

- Neimo, Leo, ed. (1999). "13". Papermaking Chemistry. Papermaking Science and Technology. 4. Helsinki, Finland: Fapet OY. pp. 289–301. ISBN 952-5216-04-7

- Houssni El-Saied, Altaf H. Basta and Mona M. Abdou. "Permanence of Paper 1: Problems and Permanency of Alum-Rosin Sized Paper Sheets from Wood Pulp." Restaurator: International journal for the Preservation of Library and Archival Material, 19, no. 3 (1998): 155-171

- V. Daniels and J. Kosek, . "Studies on the Washing of Paper, Part 1: The Influence of Wetting on the Washing Rate." Restaurator: International journal for the Preservation of Library and Archival Material, 25, no. 2 (2004): 81

- Burnham, Dorothy K. (1980). Warp and Weft: A Textile Terminology. Royal Ontario Museum. ISBN 0-88854-256-9

Permissions

All chapters in this book are published with permission under the Creative Commons Attribution Share Alike License or equivalent. Every chapter published in this book has been scrutinized by our experts. Their significance has been extensively debated. The topics covered herein carry significant information for a comprehensive understanding. They may even be implemented as practical applications or may be referred to as a beginning point for further studies.

We would like to thank the editorial team for lending their expertise to make the book truly unique. They have played a crucial role in the development of this book. Without their invaluable contributions this book wouldn't have been possible. They have made vital efforts to compile up to date information on the varied aspects of this subject to make this book a valuable addition to the collection of many professionals and students.

This book was conceptualized with the vision of imparting up-to-date and integrated information in this field. To ensure the same, a matchless editorial board was set up. Every individual on the board went through rigorous rounds of assessment to prove their worth. After which they invested a large part of their time researching and compiling the most relevant data for our readers.

The editorial board has been involved in producing this book since its inception. They have spent rigorous hours researching and exploring the diverse topics which have resulted in the successful publishing of this book. They have passed on their knowledge of decades through this book. To expedite this challenging task, the publisher supported the team at every step. A small team of assistant editors was also appointed to further simplify the editing procedure and attain best results for the readers.

Apart from the editorial board, the designing team has also invested a significant amount of their time in understanding the subject and creating the most relevant covers. They scrutinized every image to scout for the most suitable representation of the subject and create an appropriate cover for the book.

The publishing team has been an ardent support to the editorial, designing and production team. Their endless efforts to recruit the best for this project, has resulted in the accomplishment of this book. They are a veteran in the field of academics and their pool of knowledge is as vast as their experience in printing. Their expertise and guidance has proved useful at every step. Their uncompromising quality standards have made this book an exceptional effort. Their encouragement from time to time has been an inspiration for everyone.

The publisher and the editorial board hope that this book will prove to be a valuable piece of knowledge for students, practitioners and scholars across the globe.

Index